Praise for *The Fruit Forager's Companion*

"Sara Bir's voice is quirky, informed, and fresh. *The Fruit Forager's Companion* will push any soul who is interested in foraging into the curious world of fruits, which are every bit as interesting as the vegetable members of the plant world. I just hope that she refrains from lifting my quince should she ever walk down our lane—I adore them, too! Which is to say that you want someone with passion and appetite to lead you on a foraging quest, and Sara Bir has plenty of both."

—**DEBORAH MADISON**, author of *Vegetable Literacy* and *In My Kitchen*

"Lyrically written and eminently useful, *The Fruit Forager's Companion* is a welcome addition to the library of anyone interested in either preserving their own fruit harvest or seeking out new, exciting flavors that are literally growing on trees—often next door!"

—**HANK SHAW**, wild foods expert; author of the James Beard Award-winning website Hunter Angler Gardener Cook

"For fruit lovers the whole world is a culinary theme park; this book is your permanent admission pass. Let Sara Bir guide you to the untamed flavors of wild, feral, and neglected fruits—from back alleys and brushy waysides to city hedges and deep woods. After you find some brand new delicacy right in your own neighborhood, follow one of Sara's luscious recipes, and invite Mom over for dinner."

—**SAMUEL THAYER**, author of *Incredible Wild Edibles* and *The Forager's Harvest*

"Once you notice the wild fruit growing all around you, the world becomes a landscape of culinary abundance, and Sara Bir's *The Fruit Forager's Companion* is a thoughtful guide to appreciating those foraged and gleaned fruits. Filled with Bir's distinctive humor, the book, like stumbling upon a patch of black raspberries, is also fun! Bir's respect for ingredients—those berries, apples, and pawpaws you've worked hard to pick—ensures the recipes accentuate each fruit's unique flavor. With Bir's guidance, your fruit-forward concoctions will be as transformative as the moment you discovered all those wild fruits were edible in the first place."

—**ANDREW MOORE**, author of *Pawpaw: In Search of America's Forgotten Fruit*

"Among reassuringly familiar fruit like neighborhood apples, lemons, and plums, chef-turned-forager Sara Bir also offers curious novice foragers more adventurous fare: invasive autumn olives and barberries, native chokeberries, Oregon grapes, pawpaws, and spicebush. The pages of *The Fruit Forager's Companion* help you to identify, collect, and use the fruits of your forages. The author's eloquent introduction tells you why you should."

—**MARIE VILJOEN**, author of *Forage, Harvest, Feast*

"With *The Fruit Forager's Companion*, Sara Bir provides not only a guide to foraging, but a manifesto for conscious living and a challenge to seek out the unknown. With creative recipes, thoughtful writing, and a wealth of expertise, she encourages us to explore in the kitchen as well as outside, inspiring the reader to create a better connection to where they live and to celebrate the local bounties that may have otherwise gone unnoticed."

—**ANNA BRONES**, author of *Fika: The Art of the Swedish Coffee Break*; founder and publisher, *Comestible*

"*The Fruit Forager's Companion* will be beloved by all those who travel through life scanning trees and shrubs for neighborhood fruit. Sara Bir has created something that is half foraging memoir and half cookbook, and it is utterly delightful in its totality."

—**MARISA McCLELLAN**, author of *Naturally Sweet Food in Jars*

"Lyrical and practical, introspective and funny, *The Fruit Forager's Companion* inspires us to put on comfy shoes and head out into the local landscape with curiosity, confidence, and joy. Sara Bir knows that sweet, ripe treasures await us, from crab apples and blackberries to pawpaws, wild grapes, figs, and quince. This book offers fascinating entries on more than forty fruits and a hundred recipes for chutneys, soups, cordials, fools, and more. Bir's knowledge, wit, and enthusiasm guide us outdoors for fruit foraging expeditions, and back home again to transform the seasonal fruits we've gathered into good things to eat and share."

—**NANCIE McDERMOTT**, author of *Fruit* and *Southern Pies*

"Sara Bir's common sense approach to foraging, along with an impish humor, make for a delightful, nourishing, very practical, and very human read. *The Fruit Forager's Companion* is a book about love, community, and the abundance nature offers to us all if we have the eyes to see and the heart to hear, all revealed through the simple, graceful acts of picking, preparing, and sharing wild fruits with our loved ones and community."

—**ROBIN HARFORD**, founder of the website Eatweeds

"The response to the lament 'I'm hungry' should not be, 'Look in the fridge and see what you can find,' but, 'Take a walk and see what's there.' Sara Bir's book provides a road map to wild and abandoned plants laden with food. As you read and explore with Bir you will be rewarded with the joy of discovery and often a satiated appetite. Perhaps you will even find a dead-ripe mulberry and have an ecstatic taste experience."

—**TOM BURFORD**, pomologist, historian, and author of *Apples of North America*

"Once, we were foragers. Sara Bir says we can be again. She reveals the wealth of fruit waiting to be picked in wild and not-so-wild places, and she shows how foraging benefits the mind and body even if the forager returns empty-handed. *The Fruit Forager's Companion* is more than just a guide to finding, gathering, preserving, and cooking. It is a meditation on modern life and how to find meaning in Nature's larder."

—**MIKE SHANAHAN**, author of *Gods, Wasps and Stranglers*

"Foraging for fruit is all about noticing and making your move when things become ripe. In *The Fruit Forager's Companion*, Sara Bir moves from city sidewalks to deep woods with a botanist's eye and a chef's skill. She boils, reduces, ferments, dips into history, and seasons with memoir; she gets in there and shouts wild flavors out with heat, sweet, salt, and vinegar. Let Bir's inventive recipes and sheer derring-do pull you into the woods and make you a forager: a sampler of the best things in life, most of them free."

—**JULIE ZICKEFOOSE**, author of *Baby Birds*, *The Bluebird Effect*, and *Letters from Eden*

THE FRUIT FORAGER'S COMPANION

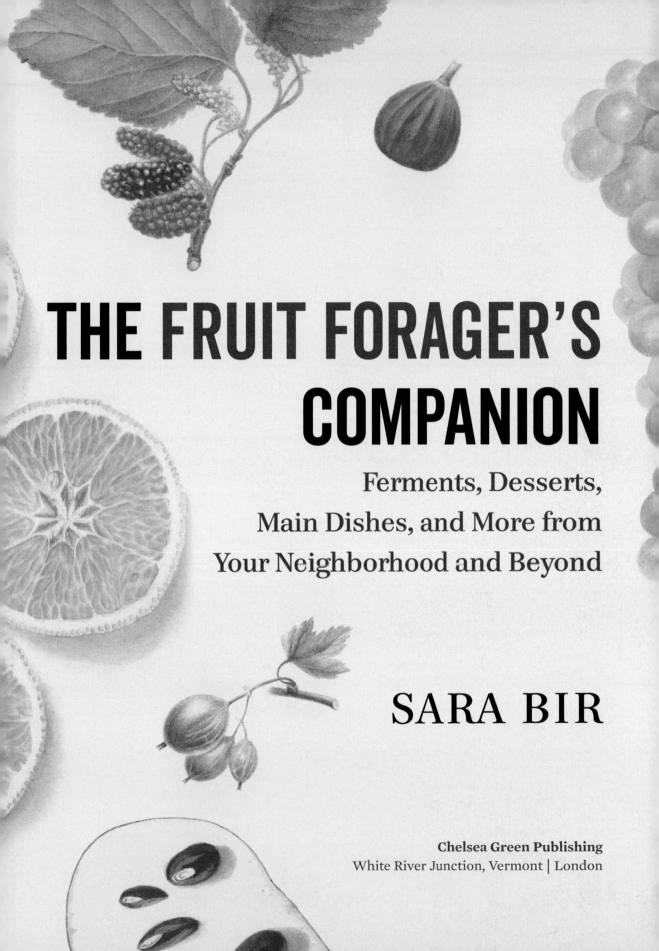

THE FRUIT FORAGER'S COMPANION

Ferments, Desserts,
Main Dishes, and More from
Your Neighborhood and Beyond

SARA BIR

Chelsea Green Publishing
White River Junction, Vermont | London

green press INITIATIVE

Copyright © 2018 by Sara Bir.
All rights reserved.

No part of this book may be
transmitted or reproduced in
any form by any means without
permission in writing from
the publisher.

Project Manager: Patricia Stone
Editor: Michael Metivier
Copy Editor: Laura Jorstad
Proofreader: Eileen M. Clawson
Indexer: Linda Hallinger
Designer: Melissa Jacobson

Printed in
the United States of America.
First printing May, 2018.
10 9 8 7 6 5 4 3 2 1   18 19 20 21 22

Chelsea Green Publishing is committed to preserving
ancient forests and natural resources. We elected to
print this title on paper containing at least 10% postconsumer recycled paper, processed chlorine-free. As a
result, for this printing, we have saved:

20 Trees (40′ tall and 6-8″ diameter)
9,280 Gallons of Wastewater
8 million BTUs Total Energy
621 Pounds of Solid Waste
1,711 Pounds of Greenhouse Gases

Chelsea Green Publishing made this paper choice
because we are a member of the Green Press Initiative,
a nonprofit program dedicated to supporting authors,
publishers, and suppliers in their efforts to reduce their
use of fiber obtained from endangered forests. For more
information, visit www.greenpressinitiative.org.

Environmental impact estimates were made using the
Environmental Defense Paper Calculator. For more
information visit: www.papercalculator.org.

Our Commitment to Green Publishing

Chelsea Green sees publishing as a tool for cultural change and ecological stewardship. We strive
to align our book manufacturing practices with our editorial mission and to reduce the impact of
our business enterprise in the environment. We print our books and catalogs on chlorine-free
recycled paper, using vegetable-based inks whenever possible. This book may cost slightly more
because it was printed on paper that contains recycled fiber, and we hope you'll agree that it's
worth it. Chelsea Green is a member of the Green Press Initiative (www.greenpressinitiative.org),
a nonprofit coalition of publishers, manufacturers, and authors working to protect the world's
endangered forests and conserve natural resources. *The Fruit Forager's Companion* was printed on
paper supplied by QuadGraphics that contains at least 10% postconsumer recycled fiber.

Library of Congress Cataloging-in-Publication Data

Names: Bir, Sara, 1976– author.
Title: The fruit forager's companion : ferments, desserts, main dishes, and more from your
   neighborhood and beyond / Sara Bir.
Description: White River Junction, Vermont : Chelsea Green Publishing, [2018] | Includes index.
Identifiers: LCCN 2017058242| ISBN 9781603587167 (pbk.) | ISBN 9781603587174 (ebook)
Subjects: LCSH: Cooking (Fruit) | Forage plants.
Classification: LCC TX811 .B585 2018 | DDC 641.6/4—dc23
LC record available at https://lccn.loc.gov/2017058242

Chelsea Green Publishing
85 North Main Street, Suite 120
White River Junction, VT 05001
(802) 295-6300
www.chelseagreen.com

FSC
www.fsc.org

MIX
Paper from
responsible sources
FSC® C084269

O bitter black walnuts of this parachuted earth!
O gongbirds and appleflocks! The friend
puts her hand on your shoulder. You find
a higher power when you look.

—MATT HART, from "The Friend"

CONTENTS

INTRODUCTION

I'M NOT A FORAGING EXPERT, and I'm not a botanist; I'm a physically restless chef who likes plants. I'm attracted to fruit because plants developed their fruit to be attractive to animals, and I am a curious animal following their lead. I don't generate any income from foraging, and the food I bring home represents a tiny sliver of my annual diet. But even if I come back from an outing empty-handed—which is most of the time—I return enriched, because there's always something new to see. I don't just gather food. I gather observations.

This book is about fruit and foraging, but it's really about walking and noticing, activities that create a lens through which the world around us comes into clearer focus. Walking is a huge part of what makes us human. It's how we get from one place to another, even if the distance is only from the front door to a car door. On foot we absorb information differently. Our contact with everyday living things is more intimate. We can hear leaves rustle. We can hear insects hum. We can look fellow passersby in the eyes and smile.

We can see things grow.

The world offers us unspeakable atrocities and moments of unfathomable beauty, sometimes in the same event. But in between those two extremes is the stuff of life. That middle space is defined but vast. The middle is not one thing, the way a medium-sized soft drink is one thing or Mama Bear's bowl of oatmeal is one thing. The middle is almost everything.

This middle place is our Eden. Eden is not static, the way popular culture depicts Adam and Eve hanging out in blissful ignorance with no concept of time or personal needs or desires. That would be impossibly boring, and therefore hellacious. Eden is dynamic. There's give and take, a constant series of millions of small lives reacting to external circumstances and in turn setting off new reactions.

Foraging for fruit is not the only way to engage with our Eden, but its sensual aspects are, by design, difficult to resist. There is sweetness, color, fragrance, bitterness, juiciness, sourness, lusciousness. Blooms and pollinators. Producers and consumers. Ends and beginnings. There is looking and finding. Foraging is our journey in Eden.

My daily walks make up my foraging practice, the way other people have a yoga practice. Most of my foraging practice is prosaic stuff, like walking my daughter the few blocks to her elementary school in the morning. It's the same route over and over again. I take comfort in the routine. Sometimes I'll modify my route back home so I can see the progress of specific fruit trees, depending on the season: the persimmon tree in the fall, the plum tree in the summer. It's like stopping in on friends.

I came into this world with a heartier dose of existential dread than the average person. I'm not sure why—I have a loving family and a stable home—but unresolvable issues about the human condition have always plagued me. My daily walks are more than a mechanism for better circulation and getting fresh air in my lungs. If I don't get out of the house at least once a day, I lose it. My temper flares up, my energy level plummets, and I feel too much inside of myself.

The existential dread is an emotional cancer, and my foraging practice is a distraction for those overeager white blood cells that cause trouble. I channel that destructive energy into something constructive. Worst-case scenario, I get exercise and wave at some neighbors; best-case scenario, I bring home something interesting to eat.

Foraging has also made me less egocentric. Instead of thinking of a route as an obstacle to endure—a thing standing between me and a location where I will accomplish a necessary task, like driving to the grocery store or walking to the library—I think of each journey as a holistic experience. My neighborhood was not erected specifically for my convenience. It is a settlement, just as anthills and squirrel nests are settlements on the same land as the streets and sidewalks I use.

This is at odds with one of the fundamental ways we define our natural resources. Is our planet a global convenience store of energy and water and raw material for us to help ourselves to without reservation? From that viewpoint, everything exists for people, like props and scenery on the set of a play.

I've been grappling with the concept of a seven-day creation with mankind at its apogee ever since a kindly Sunday school teacher first presented the lesson to our young class. It never clicked for me. In my own lifetime, I've noticed changes our consumption habits have initiated. I gaze into the woods and see a lot more of some plants and a lot less of others. I hear the dreaded racket of leaf blowers drown out the backwash of other, gentler sounds. I see deer move closer and closer to our homes, because we've moved closer and closer to theirs.

I'm complicit in all this, of course. I cram garbage into nonbiodegradable plastic bags and shove the filled bags into our garbage can to eventually be buried in the earth. I book flights and buy frozen pizzas imported from Italy and grapes grown in Chile. I enjoy the comfort our first-world status affords us.

But I know it has its limits. It can't last, and no one making money from the perpetuation of that lifestyle is happy to admit it. The powerlessness I feel as a participant in a cobbled-together system that ruins what we hold most dear overwhelms me, and sometimes I get numb and push it all to the back of my head so I can keep on going.

Foraging is, to me, a daily, small act of civil disobedience. Simply looking for wild stuff to eat is a way of flipping the bird to our industrialized food system. It's a way of asserting our intellectual curiosity and reclaiming our natural role as humans on this planet, something we've allowed our modern lifestyle to strip from us. Foraging restores the balance a bit. Even if it's just a handful of blackberries or a few fallen apples from a neighborhood tree—these things still count. The magic of foraging is ultimately about the quality of time you spend interacting with the world around you, not the quantity of edibles you haul home.

I started foraging because I am nosy and cheap. I like free things, and I like to look at property that does not belong to me and imagine what goes on there. Do this long enough and you will notice different types of fruit growing, some of it free for the taking. I noticed the fruit, and I took it.

When I lived in California and was just embarking on my career as a chef and writer, my foraging was sporadic and casual. Fruit just happens in California; you'd have to put an effort into avoiding it. I didn't care as much about the trees as what they produced, and I mostly remember receiving sacks of Meyer lemons and Hachiya persimmons. Two plum trees in the backyard of a shared rental house dropped buckets of teeny, squishy, giant-pitted plums that I'd bake in coffee cakes.

I moved too often to undertake gardening, so I relied on other people's plants. I liked the adrenaline rush of unexpected fruit windfalls, which sent me into the same elated states as a good score at the record store or sighting barely damaged furniture on the curb. Some people's role is to drag home what is unloved and give it a place to belong.

Belonging is elusive. It comes and goes. I grew up in southeast Ohio, where Midwest meets Appalachia. It was fine enough, but always flummoxed me because its midwestern understatement and Appalachian self-reliance bled together to form a specific yet unspoken regional identity I couldn't summarize in one handy sentence to use as a reply when people asked me where I was from. It's appropriate that I didn't see my first pawpaw—to me the Sasquatch of fruits—until after I'd lived for years away from where they grew, when I'd defined myself as someone who left and would never return. But I eventually grew disenchanted with the easygoing magic of West Coast life and struggled with profound stirrings to move back to my small hometown; wanting to settle there again meant I'd had no idea who I was all along. But we packed up and did it—I had a husband and a young daughter by then—and the adjustment was trying. There were ugly arguments and job search woes.

To work through the stress I'd created for us, I took to the woods and did some heavy-duty forest bathing. I wasn't even looking for pawpaws when I found my first. I was walking on a trail in the woods, and one lay splayed open smack in the middle of the path in front of me, displaying its saffron-colored guts.

I don't believe in fate, but I do believe in the magic of being receptive to the right thing in the right moment, and when I encountered that pawpaw, that was it. I went back and looked for more. They smelled and tasted like nothing I'd ever encountered in Ohio, though they'd been growing there for millennia. I fell for their amazingly tropical flavor and intensely perfumed flesh. I started hauling them home and have been at it ever since. It's like that pawpaw found me, not vice versa.

When my pawpaw obsession was still new, my brother thru-hiked the Pacific Crest Trail. It's easily one of the definitive episodes of his life. As he shared crazy stories and amazing, scenic photos, I wasn't exactly envious, but I did yearn for the days when I'd summited gnarly peaks and touched the face of a rawer, seemingly more majestic incarnation of nature. There I was, back in Ohio, living amid so many things that had barely changed since I was a kid. No mountains, no canyons, no ocean coastline.

But what had changed was me. I saw my hometown differently. Instead of having nothing to offer me, it had everything to offer. I had a daughter and a husband and a dog, and it was the right place for us. I had to be open to wonders on a smaller scale if I wanted to have any adventure in my life. I had to allow my surroundings, rather than my expectations, to redefine what was epic.

So while my brother made his way north through the Laguna Mountains all the way to the Canadian border, I undertook my own ambitious journey. I hiked over the same stretches of trails day after day. The trailhead was right behind where my daughter went to preschool, so I'd drop her off and set out on my little adventure for the day. I liked how intimately I got to know the curves and dips in the trail, where to avoid roots or poison ivy.

I noticed new things every time. I watched as the pawpaw trees budded, then bloomed, then sent out their long, gently tapered leaves. I saw the fruit set, and grow, and then I snatched up pawpaw after pawpaw as they became ripe. When fall came I saw those leaves turn gold and eventually shrivel up, then fall to the ground and leave behind short, slender trunks and branches. I noticed all kinds of insects and fungus and moss.

Pawpaws were my gateway drug of foraged foods. Every year I see or hear about new things, and my repertoire expands: crab apples, mulberries, sumac, autumn olives, spicebush. Off the trails I see what's growing on trees in yards and gather stray apples and persimmons off the sidewalk. I learn about fruits that don't grow here, and I try to look for them if I happen to be elsewhere. But setting out to look for fruit isn't as fun as leaving the house and being open to what you happen to come across. If you are receptive, what's seemingly insignificant can become one of the most significant parts of your life. Who knows where fruit could lead you in the future? Let it open you up to follow your instincts and see the magic that happens because of it.

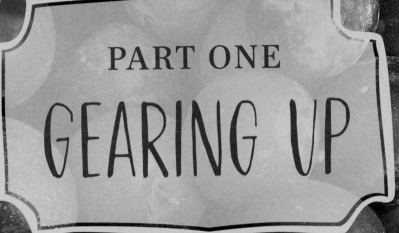

PART ONE

GEARING UP

FORAGING AND YOU

WE LIVE IN AN ERA of insane abundance. This hits me every time I haul a bag of stuffed animals to the Goodwill. About once every season we do a purge of my daughter's bedroom, and all of the books and toys and clothing she has grown out of go to be of use to another family (or so we hope). I hate shopping, and so does my daughter; most of her toys are hand-me-downs or gifts or yard sale finds. But good God, to think that one person barely over a yard tall can amass so much stuff *without even obtaining it directly.*

I enjoy sewing, and if you've ever made a stuffed animal yourself, you know it takes work and skill. This is something I try to impress upon my daughter when she handles an object carelessly. "Someone made that," I say. "When you cut holes in your shirt, you're disrespecting the work someone put into creating it." This has zero effect on her (she still cuts holes in her shirts!), but I feel like the basic principle applies in even greater measure to the food we eat. The US Department of Agriculture estimates that roughly 30 to 40 percent of the American food supply ultimately goes into the garbage. It takes natural resources like water, land, and energy to grow food, but it also takes human energy, and to be wasteful devalues the dirty and hard work of growing food. If you've never watered a garden day in and day out or spent hours crouched over picking strawberries under a blazing sun, that hard work is easy to take for granted.

Foraging for fruit is not going to save the world, but it's an act that can help make us better people. It enriches lives, gets us out of doors, and ignites our curiosity. It connects us with subtle shifts in the seasons and

brings us closer to the people and animals who share wild and settled spaces with us. It is hard and dirty work with a payoff beyond calories: a step toward a more sustainable food system and a lifestyle that nurtures rather than harms the land we depend upon. It's walking the walk.

Foraging reminds us where we came from. Human societies by nature evolve and change, but the pace of those changes feels swifter than ever. Our deeply ingrained instincts have new impulses to keep them company: social media tapping our attention, an array of electric lights disrupting natural sleep patterns, pantries bursting with refined grains and artificial flavors.

Before the Agricultural Revolution and the domestication of plants, humans obtained food from foraging and hunting. Some anthropologists mark the beginning of agriculture as the first nail in the coffin of humanity. Growing food and tending to livestock gave way to a more settled way of life, and thus permanent villages, towns, and cities. Pollution, overpopulation, disease, and the pilfering of natural resources followed. Humans innovate by nature, and those innovations inevitably alter the planet we inhabit. Anthropologist Jared Diamond once famously argued that agriculture was "the worst mistake in the history of the human race," noting studies of contemporary hunter-gatherer tribes, who have ample leisure time, varied diets, and no dependence on the vulnerabilities of monoculture.

So here we are today, standing on the shoulders of scores of human generations before us. I won't say it's for better or for worse, because I was not alive twelve thousand years ago, and I really enjoy things like movies and synthesizers and preventive dentistry. I also like olive oil and chocolate, and neither is made with things that grow where I live, but I'm not going to stop consuming them. We are human animals, but also products of our time, and to deny that is shortsighted and elitist. To be human is to constantly defend ourselves against an idealized version of what we previously used to be. We hate what we have accomplished, but we also relish it.

What Counts as Foraging?

Foraging can happen anywhere. The succinct dictionary definition of *forage* is "to search for food or provisions." Foraging is not limited to the wilderness, because it can't be. What true wilderness remains in our developed world?

Your foraging happens in the new wilderness, the place where the natural world and the developed world meet, the *Homo sapiens* Subduction Zone. Nature is dynamic just as humans are dynamic. The friction between the two forces is the hidden, relentless magic of time.

My definition of foraging is quite flexible. I just think of it as gathering food from unexpected sources, but before the gathering comes the searching. Foraging begins with looking. Plenty of times I've gone out foraging and returned empty-handed, but I count each outing as a success, because I return with an experience. In looking for food plants to gather, I observe how the plants have changed since I saw them last, and I come away with knowledge. I notice the groundhogs and crows that are out that day, or the sound of the pressure shifting in the sky, or my neighbor out walking his dog. I might notice very little, because I have something intense going on in my head and I need to work it out with footsteps. But I still come back with something more than I started out with. That's foraging.

Foraging and gleaning are spiritual kin. I differentiate between the two this way: Gleaning is making use of salvageable food that's been harvested and discarded. Foraging is harvesting from wild plants, or plants whose primary purpose is no longer to produce food. In this book, picking cherries from your neighbors' yard at their invitation counts as foraging rather than gleaning.

Ultimately, such distinctions are splitting hairs. It's actions that matter, not terminology. Gathering food from unexpected sources offers unknown pleasures. This is your invitation to experience them.

Romanticizing nature is easy when you don't have a lot of interface with it, but even that's a fallacy, because nature happens in spite of us. We pave acres, and nature springs up in the cracks and along the periphery. Nature includes us. Foraging does not exist in a pristine bubble, which is good news for you, because it means a lot more turf is fair game. In the context of this book, foraging is defined as legally obtaining food that grows in public and private places. That's it. You don't have to be lost in the wilderness for foraging to be a significant part of your life. It's a practice and an art, one that continues every time you set foot outside your door.

The friction between Modern Humans and Traditional Humans—I think of it as the *Homo sapiens* Subduction Zone—is a fascinating place. That's where the action is, philosophically and emotionally. Foraging puts you right there. It's where wild meets cultivated and native meets invasive. Some plants are beneficial and some are harmful, but ultimately all plants have the potential to be either. Is one species overcrowding other, more desirable plants? Did you eat the wrong part, or too much of an otherwise helpful part? Does it taste wonderful but trigger an allergic reaction? It all depends on where you are, what you need at the time, and what you happen to know. Plants are only good or evil because of how we feel about them at the moment. Considering them anew brings your prejudices to the fore and gives the boundaries of your judgment a workout. It makes the big picture bigger.

You can think of foraging as prayer, meditation, gleaning, theft, survival, exercise, or fate. Take your pick, because it's all up to you, and that's the point. You choose your own adventure, harvesting the sweet serendipity along the way.

What's Different About Foraging for Fruit?

Fruit is the currency of memory. People have powerful emotions about fruits they picked and ate in their formative years, emotions they don't seem to form with the same intensity for other foods. If you grew up where mangoes grow and then moved away to a colder part of the world where they do not, the lack of mangoes is almost like nutrients missing from the cells of your cultural identity.

People seek out the fruits of their youth as they grow older. Mention old-time fruits like mayhaws and pawpaws around the right crowd (or right generation) and they will light up. It feels good to talk about something you love without having to explain yourself. When you love weird fruit, there's often a lot of explaining to do to the uninitiated.

Fruits have distinct forms, and to the inexperienced eye they are easier to identify and less abstract than other plant features, like tree bark or leaf arrangement. Those things are most often green and brown; fruit is more colorful, and in many cases something we've seen a thousand times in our lunch boxes or on produce stands. There's a familiarity there, a comfort.

Beyond the comfort, fruits are novel. Oftentimes they are the wackiest-looking part of the plant. Fruit wants to be noticed. We grow fruits in yards and parks because the plants that produce them offer us beauty (and, at times, shade and a natural barrier). That sweet nodule of pulp encasing seeds is just a bonus.

Community and Respect

Nature is our all-you-can-eat buffet, but with some caveats. Don't let carelessness ruin it for other people. Forage legally and mindfully, on both public and private land.

My foraging habit has made me a more neighborly person. When I first started picking fruit from strangers' trees, I was too shy to ask them if it was okay, and I did it in a stealthy manner, hoping no one would notice. Perhaps I was satisfying a nagging criminal urge, but mostly I was cowardly.

I've finally overcome my shyness, and in all my years of knocking on doors to ask if it's okay to gather fruit, no one has ever said no. Sometimes I get this befuddled reaction, almost like they can't get why I even asked, but ultimately it's the right thing to do.

So much of our lives happen in isolation—we live in rooms in houses on streets in a grid, and we work in offices partitioned by feeble walls. I hardly know who my neighbors are, and they hardly know me, though some of them now know me as the crazy lady who collects japonica quinces from their shrubs—a reputation I'm totally okay with. I appreciate the opportunities for interaction with strangers that my foraging habit has presented me. It breaks up the solitary quest of the lone forager and circles us back to the human communities we are part of.

One summer I discovered that our local Walmart had fruit trees growing on the outskirts of its parking lot: peaches, pears, and apples. Naturally I went to investigate, and it turned out most of the trees—there were a dozen or so—had plaques or wooden markers mounted on the ground beneath them in memoriam to employees who had died from chronic illness.

This little fact instantly changed my relationship to those trees. Initially I'd gone over because I was curious, but then I became pensive. What did Walmart do with the fruits of those trees? Did some of the families come by every year to harvest the fruits of the tree honoring their loved one? Did current employees go out and pick any as snacks? Maybe Walmart gathered all of the ripe fruit in one swoop and donated it to a food pantry.

None of those scenarios seemed likely, and I didn't care enough about the true answer to bother with asking management. My best guess was the fruit fell to the ground and then rotted, as is the case with so many ornamental fruit trees. I noticed a homeless fellow with a shopping cart full of belongings who was resting in the shade of one of the apple trees, and I thought I might ask if he knew, but it occurred to me he might take offense at my implication that he possibly scrounged for fruit because he was homeless. My fruit scrounging, meanwhile, was purely for amusement. I picked a hard little peach off the grass under one of the peach trees—peaches don't do well where I live—and decided to partake of the Walmart fruit on a visual level only.

Assessing the legality and decency of a foraging scenario operates on a case-by-case basis in your brain and in your gut. I know from antics back in my high school days that I am very good at sneaking around doing forbidden but harmless stuff and not getting noticed. The eighteen-year-old Sara would have been all over those Walmart fruit trees, probably sometime after midnight. With the judgment of a seasoned adult, I decided asking for permission would take too much effort, and it was best to move along to the next opportunity. What you might do in a similar situation is up to you, though if you harbor any doubts, listen to them and walk away.

Foraging is illegal in some national parks and many government-owned state and regional lands. Some national forests allow the harvesting of specific wild foods but may require a permit or charge picking fees. Is there a difference between a handful of trailside berries and the hauling out of bushels of berries?

No matter where you pick, a good rule of thumb is to take some for yourself, leave some for the wildlife, and leave some for the next forager. If you have express permission from the plant's owner to clean house, though, then by all means do so.

Foraging and Safety

"There are old mushroomers and bold mushroomers, but no old bold mushroomers," an old saying goes. It applies to foraging for any edible.

Coveting

"How do we learn to covet, Clarice?" This is something Hannibal Lecter asks agent-in-training Clarice Starling when she goes to question him in prison about killings she is investigating in the 1991 film *The Silence of the Lambs*. Impatient, Lecter answers himself: "We begin by coveting what we see every day."

I didn't really get what this meant for years, not until I started foraging in the same spots over and over again. *We covet what we see every day.* I coveted the quince growing in a yard I often passed as I walked my dog. I coveted the black raspberries growing in a clearing in the woods I ran through every week. I coveted the figs on the tree I drove by all the time.

Repeated exposure gives us a sense of attachment that a onetime glimpse does not. This extends far beyond fruit. I coveted a pair of red cowboy boots I saw in a shop window on my way to Penn Station at the close of every day when I worked in Midtown Manhattan. I coveted a 1957 Chevy Bel Air parked in a parking lot I rode my bike past a lot. I seriously had a crush on that car. It was for sale with an asking price of $5,000. I happily fantasized about buying it, even though I didn't have the money or mechanical aptitude for it, plus I don't like to drive.

The serial killer Hannibal Lecter alluded to in his coveting spiel coveted the bodies of women. Fruit is so much more wholesome in comparison, isn't it? Increasingly, I covet seeing the fruit more than getting the fruit itself; I covet the sense of purpose. Crushes and coveting can lead to irrational behavior, but they can also be intoxicating. You might as well enjoy the ride.

Eat the wrong thing at the wrong time or in the wrong amount and you could get sick or die. Such warnings are alarming and discouraging, though I figure if you've made it this far, you're in it to win it. Here's how to forage safely.

There is a big difference between fear and caution. If you are cautious, there's no need to be afraid of foraged foods. But don't be lax about your homework or your information sources, because that's when you can get into trouble.

Never taste an unfamiliar wild plant until you have positively identified it. This may sound like overkill, but I like to cross-reference a few field guides or websites before making conclusions. Make sure your sources are credible. It's all too easy for a person to call themselves a foraging expert, but I've noticed a lot of misinformation in foraging videos and blog posts online. I've included a list of my favorite print and online resources for plant identification in the resources section at the end of this book.

Actual people you meet face-to-face are invaluable sources, but unless you have already established trust with someone, don't assume they know what they are talking about. Sometimes even the owners of the property you are foraging on might mistake the identity of a plant. Also, if I share a photo of a plant I'm unsure of on social media, I defer to the comments of people I know have years of experience out in the field.

After you've identified a plant and confirmed it's in the clear, there's still homework to do. This is great news, fellow nerds! Plant homework is what we live for. Understanding what parts of the plant are safe to eat, and the manner of preparation needed to render them so, is the second half of your research.

In 1983, in California's Monterey County, a group of people at a retreat crushed some stems and leaves in the vat full of elderberries they'd picked. They all drank the juice, and on the following day eight of them were flown by helicopter to the hospital with severe nausea, vomiting, and abdominal cramps. That whole episode could have been avoided if they hadn't plunged so overenthusiastically into preparing and consuming foods they were underinformed about. Elderberry leaves and stems are toxic, and raw elderberries themselves aren't agreeable except in small quantities.

Some forageables are easy to identify and safe to eat, but with caveats. Let's use pokeweed as an extreme example. Humans can technically consume the flesh or juice of pokeweed berries without croaking, but their seeds are toxic, while pokeweed leaves need to be blanched a number of times before consumption in order to purge them of their toxins. I'm not terrified of pokeweed—it's grows all over the place and does not lunge out and attack people—but I figure if other forageables grow nearby and require less effort to eat with bigger rewards, why not just focus on those? This is caution, not fear.

Some fruits, like ginkgo nuts and mayapples, are safe to eat but not advisable to overindulge in. When you're trying a foraged plant for the first time, sample a small amount, just to be safe.

Another safety precaution: Don't get yourself into a pickle just to pick fruit that's out of reach. There are muddy slopes and spiny brambles and poison ivy to contend with. It's smart to wear long pants when foraging in scrubby or woodsy areas so you avoid scratches and abrasions (I'm going to be honest: I wear sundresses and sandals in the summer and break this rule often). There are a few times I've climbed too high or reached too far, risking a twisted ankle or worse. Especially if you are foraging alone, err on the side of caution. You don't want to hike out on a gimpy leg just so you could get that giant gorgeous pawpaw.

Lots of edibles grow on the side of the road, but they could be contaminated with residue and exhaust. Most fruit grows well above the ground, from branches and bushes, so chemical contamination isn't as much of an issue as it is with foraging greens, but it's still important to be mindful. Where is the plant growing—on a Superfund site? Near any underground storage tanks?

Bad judgment is just as dangerous as physical hazards. I've had a few foraging-related mishaps, most notably a time I set out to collect black cherries from a heavily laden tree I'd been eyeing. It was on a slope in a tricky spot, all the more reason to conquer it. I climbed a fence and spread a tarp under its branches, then whacked at them with a long stick to jostle them free so they'd land on the tarp. Just as the cherries began to rain down promisingly, I felt sharp jabs on my legs and arms: I was standing on a yellow jacket nest. I skedaddled out of there with six stings and considered the situation from a distance. I was still hell-bent on getting those cherries, so I gingerly retrieved my tarp once the yellow jackets settled down and reluctantly called it a day, imagining the possible headline LOCAL FORAGER DIES OF STUPIDITY. No fruit windfall is worth a potential trip to the hospital, which is where I ended up—a ride in the ambulance and three hours in the ER after I passed out in the checkout line at a drugstore while waiting to purchase Benadryl.

Failing to notice an underground yellow jacket nest was not reckless, but going back there to get less than a quart of cherries after I'd been stung multiple times was. After that episode, I decided my penance was to avoid collecting anything beyond casual handfuls of black cherries for the duration of the season.

Tools for Foraging

Happily, foraging for almost anything (fruit included) requires minimal gear. You probably already have most of this stuff in your garage or basement, should you even need it.

You Are a Scientist

Did you love science class as a kid? Was it special when you got to do seemingly simple experiments, like making a compass out of a cork and a needle? Were you totally fascinated by liquid mercury? Did you burn ants with a magnifying glass out of not cruelty, but curiosity?

As we become adults, that sense of wonder ebbs. We understand more about how things work, and then we get jaded and arrogant. We get distracted. But we can still rekindle our love of discovery. Just because a plant was first discovered by humans thousands of years ago doesn't mean we have to miss out on the fun of unraveling the mystery ourselves. The everyday world is our laboratory. What happens to those leaves when the weather gets cool? What will the flowers look like in the spring? Do they smell? What kinds of animals hang out around that plant? What does the fruit taste like when it's still unripe?

Books and websites and apps can tell us this stuff, but why just be told? Maybe there's no rush. You can go out and face the world as if it's full of totally new life-forms. Collect data and analyze it. Be a third grader again. Foragers are amateur scientists.

In your kitchen you're still a scientist. Sometimes with niche fruit you *have* to be a scientist, because to execute your vision, there's no written guide. Can you make loquat juice into *pâté de fruit*? Do Asian pears make good vinegar? What happens when you dehydrate unsweetened cranberries? Now that your laboratory has just become everyplace you set foot, what will you observe next?

Sticks: My favorite tools. Sometimes you can pick up a long stick and swat at tree branches to knock down those out-of-reach fruits.

Drop cloths or tarps: These can go along with the stick. Laying down a tarp under a cherry, mulberry, or hackberry tree can save you the effort of picking the dropped fruit from the grass below.

Hats: In a pinch, I've picked berries and put them in my hat because I didn't have a bag or container with me. This technique is not recommended for people who sweat a lot.

Canvas bags: I prefer fabric to plastic because plastic bags make that awful crinkly sound, and the sounds you hear outdoors—whether it's in a busy neighborhood or way out in the woods—are all part of the experience.

Smartphone: I try to go to the trails to escape the temptations of constant e-connectivity—I think it can detract from our immediate physical surroundings—but dang, is it ever handy to have a smartphone nearby. Taking pictures, for one, is a great way to

later identify a plant you've not seen before. Some people use apps to help them identify plants.

Gloves: A matter of preference. Prickly pears are called that for a reason. Never gather them with your bare hands, because their spines are minute and splintery and will embed themselves in your palms and fingers like fiberglass. Other plants are not as hazardous for your skin but have bothersome thorns or leaves—roses when collecting rose hips, spiky leaves when collecting Oregon grape plant berries—and some lightweight gardening gloves might be a good idea for comfort. They can reduce mobility, though, and when I wear them I usually take them off before long.

Rigid containers: Delicate, easily bruised fruits like raspberries and figs do better when they lie flat. Broad-bottomed baskets, cardboard flats, or shallow plastic buckets with handles are useful. I always save pint- and quart-sized paper berry cartons when I buy produce, too—using them to harvest berries is handy, and you can more easily eyeball the quantity. Just make sure not to pile the fruit too high, because the weight of the top layers can crush the bottom layers, undoing the work you just did.

Insect repellent: Up to you. Mosquitoes don't discriminate among yards, parks, and the deep woods. In the late summer and early fall—prime foraging season where I live—mosquitoes are the most out for blood, and I spray myself up really well before heading out.

Winging It

For all my talk about connecting with nature and allowing the out of doors to slip you into another state of consciousness, I don't insist on sanctity all of the time. I listen to a lot of podcasts when I forage, though the cords of my headphones sometimes get tangled up in shrubs or branches.

I do intentionally leave my smartphone behind from time to time when I go out on walks, so I can listen better and use my brain in a different way. If I run across a nifty unfamiliar plant, I have to circle back another day, and I might forget. Sometimes I'll shove a few leaves in my pocket, and then I forget those, too, and come up with a handful of dry crumbles a few weeks later. It can take me up to a year to identify a plant at such a pace, but I don't mind. If I do remember to go back to it, I might not even look it up in a guidebook or an internet search—I tease out the identification to see how long it'll take me just by making observations or stumbling across incidental connections. During this entire time, I don't *eat* any of the plant, of course. I see how long I can wing it, because there's not always a rush. It's good to leave a few surprises for yourself when you can. When I do make an ID using this long-term method, the sense of accomplishment is as rewarding as any fruit.

Your Foraging Checklist

If you happen upon a potential foraging bonanza, go over these questions quickly in your head before you get started. This is a bit like how you're supposed to walk around your car and assess its condition for a few seconds before you drive it, or consider your site before you set up camp.

Is it safe? Are there any possible contaminants nearby—exhaust, industry, pesticides? Are there physical hazards like steep slopes, fences, or high, dangling tree branches? Are there pesky insects or rash-causing plants? Is there traffic (motorized or even bicycle) to be mindful of?

Is it legal? Whom does this land belong to? Is it public or private? Is the plant protected at all? When in doubt, pick up the phone or knock on a door and ask.

Is it for sure what I think it is? If you haven't identified the plant positively, take photos or a few leaves and fruits home instead for further research.

Foraging and Accessibility

If half of foraging is walking, then people who use canes, wheelchairs, or walkers, or have other accessibility issues, won't have the same experience, if they are able to have a foraging experience at all. Besides walking and noticing, foraging is bending, reaching, and carrying. I got a taste of this from modifying my walking routes after our aging dog began losing his stamina; he couldn't comfortably trot up hills with me, or accompany me for more than a mile without getting worn out.

So for the dog, I began taking two walks a day: one short one with him on flat ground and a longer, solo one where I scrambled around in the brush more. As I vicariously experience Scooter's declining access to the out of doors—in his prime he had boundless energy and awareness of his surroundings—I wonder if he misses purposefully moving through space as he once did. He seems happy most of the time, but there's a spark missing to him that's either a void in his dog life, or a basic winding down of his body and energy.

There are two culprits when humans who could once easily move out of doors no longer can: health and infrastructure. In developed areas where there are no sidewalks or crosswalks on busy roads, residents can't safely venture out and enjoy discovering the hidden details of their surroundings. And if you have physical mobility issues, you perhaps can't pick cherries by a mellow gravel road, much less along a route teeming with speeding cars.

Every time I set out for a walk, I consider how fortunate I am to live in an area with a mix of trails, sidewalks, and quiet streets. I feel a little uncomfortable blithely championing an activity that's not available to everyone who might like to partake. So I think of small actions to counter this inequality. A gleaning nonprofit in my town delivers food from its community gardens to seniors in low-income housing. Many of them are former gardeners whose bodies can't sustain that level of activity, and the sight of a homegrown summer tomato cheers them more than you'd think. What about bringing a small container of foraged berries to a nature lover who can't get outside any longer?

On a policy level, pay attention to the zoning laws where you live, and support elected officials who want to create accessible spaces for an active and diverse community versus rampant commercial development that cuts off access and discourages exercise.

Is it morally okay? Is this plant uncommon enough that harvesting a lot of it will make you an asshole? Ideally it's plentiful, but how much should you take? Will you be foraging or stealing?

Is it ready? This one I mess up sometimes, just because I'm so gung ho. If the fruit isn't ready to pick yet—it needs to mature or ripen longer—circle back later. Why siphon away a potentially ideal crop by bringing home inferior examples?

Fruit Maps

As a kid I was intimately familiar with every curve and dip of the street where the house I grew up in stood. It was midway up a dead-end hill. To visit my friend Susan, I walked up the hill; to visit my best friend Erin, I went down the hill. My elementary school was just down the hill, too, so at least twice a day I went in one direction or the other.

My dad, a pilot, once took me up in a two-seater airplane and buzzed over our house. I remember looking down and thinking, *That's what our street looks like?* Barely distinguishable from the surrounding streets of the neighborhood, it didn't match the image I had in my head at all. It seemed scrawny, flat.

I grew into a map lover as an adult, one of those people who can happily look at an atlas for hours. Maps fascinate me because they give off an illusion of permanence, but in truth they're a snapshot of a tiny scrap of our shifting planet. If I drew a map of my favorite foraging spots within walking distance of my house, it would be hilariously distorted when compared with an accurate aerial map. But that's the point: Our perceptions on the ground are malleable. My own imagined fruit maps—which are more like highlight reels with grape arbors and apple trees looming disproportionately large—evolve over the years as I make new discoveries, and they shift from season to season depending on what fruits are ready to pick. Which is more real: what something *feels* like, or what something actually *is*? Were we to live on the same street, your fruit map would be different from mine. It motivates me to get outside every day, looking for the magical places where any two maps intersect. You create your fruit map, but the places it leads you might not be where you expect, and that's the best thing about it.

USING THIS BOOK IN THE KITCHEN

YOU FOUND FRUIT. Now what? Figure out the best way to eat it! Crafting and executing an action plan is the second half of your foraging journey. The variability of foraged fruits will challenge you from time to time, and you will rise to the occasion because you are a problem solver.

PRESERVING FRUIT FOR LATER

The best thing about any growing thing is how ephemeral it is. Fruit, more than anything else, is a physical reminder of cycles of abundance and scarcity; when it's there, boy, is it there. When it's gone, it's gone.

Preservation and fruit go hand in hand, particularly if you are a Type A person. If you're more in the "hike and graze" camp, that's fine. Most of this book will be applicable to you, but this section is for the "gather and hoard" crowd. Every fruit has preservation methods that suit it best (in the case of pawpaws, for instance, freezing is excellent; canning, not so much).

No matter what preservation method you go with, always label and date your end product. Perhaps you know what's in it, but other people won't necessarily.

Freezing

You've probably been practicing this method without even thinking of it as preserving. Freezers are part of the fabric of our lives; we pop foods in them as an afterthought. Humans have been freezing food in cold climates for millennia, but it's only in industrial times that we've been able to freeze year-round. If you have space in your refrigerator and zip-top freezer bags, you're ready to go.

To save space and prevent freezer burn, I use a vacuum sealer when I freeze berries or chunks of fruit. The bags are more expensive than zip-tops, but I like the results, and you can reuse them multiple times. Before bagging berries, lay them in a single layer on a tray and freeze them so they'll be firm enough not to get crushed when you vacuum the air out.

Pros: Does not require much time or equipment. It's energy-intensive in some ways, but your freezer is probably running in the first place, so why not take advantage of it?

Cons: Useless if the power goes out. Changes the texture of some fruits.

Best For: Berries, slices of firm apples that will be baked later on, purees, juices, sliced peaches and apricots, cherries.

Canning

Surprisingly, canning is a relative newcomer to the food preservation scene. Consider this: Canning originated because of a cash prize offered in France during the Napoleonic wars (the government was looking for a better way to keep its troops fed). In *The Art of Fermentation*, Sandor Ellix Katz writes, "Canning, a sterilization process that revolutionized food preservation in the 19th century . . . is the diametrical opposite of fermentation." Fermentation preserves food by encouraging good bacteria to grow and create lactic acid, while canning kills all of the microorganisms. This isn't a dig at canning, which is useful and versatile and conjures up homespun images of grandmothers in calico aprons. But it's more cutting-edge than you might think.

Pros: Home-canned goods don't take up freezer or refrigerator space and can have a shelf life of years. Jars are cute and make handy last-minute gifts.

Cons: Energy-intensive. Changes the texture and flavor of some foods. Doing it makes your kitchen really hot. Requires some special equipment.

Best For: This book uses water bath canning exclusively, which is only safe for high-acid or high-sugar foods: pickles, preserves, and most juices.

Fermentation

So many cornerstones of our food culture—from bread to yogurt to vinegar to beer—are fermented, and to discover that it's possible to make this alchemy happen in your own home is empowering. Fermented foods are alive, teeming with beneficial bacteria that deter undesirable bacteria. The ferments in this book are simple wild fermentations and don't need starters or excessive monitoring, though peeking in on them as they develop is a bit like observing a new pet.

Pros: The most passive method of preservation. Results in microorganisms that promote digestive health. Easy to do in small batches. Requires little, if any, start-up equipment. Transforms the flavor and texture of foods in exciting ways.

Cons: May require a lot of space if you plan to ferment big quantities. Will extend the shelf life of many foods, but not as long as freezing or canning.

Best For: Salted-cured citrus, some relishes, fruit vinegars. (This book doesn't cover fruit wines and beers, or naturally fermented sodas, because plenty of other books discuss the subject in great detail.)

Drying

Depending on what you want to dry, there are a number of ways to go about it; outside of countertop dehydrators, there's the oven, the sun, plus the tried-and-true "hanging food from a string" approach. The

Sharpies and Masking Tape

One of the most important steps in fermentation is to label and date your jar/crock/tub of fizzing goodness. That way you can tell at a glance how long its contents have been up to whatever they've been up to. Labeling is also a way to keep your fermentation projects from getting thrown out by persnickety housemates or family members. Sometimes I even leave notes to myself on there—CHECK 7/22—so I won't forget to peek in on the progress. (I put reminders on my online calendars, too.)

I held restaurant jobs early in my working life, so labeling and dating foodstuffs is second nature to me. Sharpies and masking tape are vastly undervalued kitchen tools. I also label and date (month and year) anything I've canned—even if it's obvious at a glance through the jar what's inside—because if I give it away, maybe the recipient will forget what it is. As for the date, canned goods have a long shelf life, but sometimes as I make my way through the last few jars of a batch, I look at the date and remember what was going on in my life at the time I canned them. They're like diary entries.

I have friends who never throw anything out, including a batch of kimchi I made in 2014. I see it in their fridge every time I visit their house (it must not have been very good, or they'd have polished it off long ago). Maybe those food mementos keep happy times and faraway loved ones part of their daily lives. How sentimental you are about your preserves and ferments might have to do with how much storage space you have, but labeling and dating them will benefit you either way.

ease with which you can dry fruit at ambient temperature depends on the fruit itself and the climate of your house.

Dehydrators remove a lot of the guesswork. Not all dehydrators are made equal; it's best to invest in one with a fan and a temperature control, or your fruits will dry unevenly. The fruit you dehydrate at home might be quite different from its commercially produced counterpart. DIY dried fruits are usually darker and more leathery—or, in the case of strawberries and cranberries, oddly Styrofoam-like. I approach it all with an open mind and have fun experimenting. Refer to the guide that comes with your dehydrator; it will tell you which fruits dry best when pre-treated with ascorbic acid, or which berries need a quick dip in boiling water to crack their skins for better drying.

Pros: Time required is largely hands-off. Can be done using various methods.
Cons: A dehydrator with a large capacity takes up a lot of space and is not cheap. It can be difficult to dehydrate a bumper crop of fruit. Yield can be small, but the flavor is intense.
Best For: Orchard fruits, some berries, fruit leathers.

COOKING TOOLS AND UTENSILS

Most of the recipes in this book don't call for crazy equipment, but there are some specialized tools that will greatly ease the task of preparing foraged fruits for cooking or preservation.

Digital Scale: A digital scale saves time; it's faster than measuring ingredients by volume, and it generates fewer dirty dishes. A good one costs about $20. Do yourself a favor and buy one and use it.
Decent Stockpot: You don't need a heavy, fancypants stockpot. Actually, a stockpot that's heavy on its own will be *really* heavy when it's full of liquid. I prefer to cook acidic fruits in nonreactive pots, so it's best to have a stockpot that's enamel-lined or has a stainless-steel interior.
Pressure Cooker: Making applesauce is a snap in a pressure cooker, which is up to two-thirds faster than conventional cooking. I start all of my fruit butters off with a stint in the pressure cooker. It's also great for steamed desserts.
Crocks: Big jars are a lot cheaper, but an earthenware crock gives you that old-timey feel when you're fermenting vinegar, pickles, or kraut. If you're looking for used ones, examine them closely for cracks or chips where bacteria can hang out. I have an old crock with a micro-crack, and it took me about six batches of bad sauerkraut to figure out why it was going off.
Food Mill: Food mills are hand-operated devices that simultaneously filter and puree cooked foods, and they come in different shapes and styles. Some fruit seeds might obstruct a hand-cranked mill, which is why I prefer the conical kind that has a wooden pestle

and a tripod stand. I use it to make applesauce and process grapes, pawpaws, persimmons, and other fruits. And it's invaluable for making jam.

Electric Spice Grinder: If you collect and dry berries like spicebush, you'll need an electric spice grinder to crush them to a powder.

Mortar and Pestle: This does the work of the spice grinder, but it's hard to get spices ground as finely using one. Sometimes I prefer spices that are more coarsely ground, though.

Cheesecloth: A very clean old T-shirt can stand in for cheesecloth in a pinch, but I like to buy unbleached cheesecloth. The fineness of the weave differs in brands of cheesecloth; if yours is coarse and you are straining an infusion or cooked fruit, you'll probably want to line your strainer or colander with a double layer.

Jelly Bags and Nut Milk Bags: Jelly bags are reusable linen or nylon mesh sacks used for straining the liquid out of cooked fruits for jelly. Since they hang from either a purchased wire stand or something you rig up at home (say, tying one to the knob of a cupboard), gravity helps the liquid flow out faster and more thoroughly. You can sew your own jelly bags out of unbleached muslin. Wash them before using. I don't put them in the dryer because the fabric can shrink and make the weave tighter, which will slow down the straining process. Nut milk bags are more or less the same thing. It's totally possible to make jelly using a cheesecloth-lined colander, but jelly or nut milk bags help you extract more liquid.

Measuring Ingredients

For accuracy and speed, the most important thing you can do to improve your cooking is to get a digital scale and use it. This is quite pertinent to fruits, since volume measurements are vague and untrustworthy. And it's especially true with *foraged* fruits, because often they have more blemished spots to trim off and discard, giving you a smaller end yield than you might have from commercially grown fruit. In this book's recipes, I try to give the weight of the fruit before and after trimming as often as possible.

For ingredients like sugar and flour, use dry measuring cups and measuring spoons. Don't use a liquid measuring cup, because it's not as accurate. Likewise, you'll get better results measuring liquids in a liquid measuring cup—and fewer messes, too.

When I measure flour, I spoon it into a dry measuring cup and then level it off with a spatula or metal table knife. When I measure this way, 1 cup of flour is about 127 grams or 4.5 ounces.

Pints, Pecks, and Bushels: What Old-Timey Volume Measurements for Fruit Actually Mean

If you poke around in old cookbooks and preserving guides, the terms *bushel* and *peck* are common. These are dry volume measurements (you don't get a bushel or a peck of, say, milk). They were standardized in England in the late fifteenth century as "Winchester measure." Also, a US bushel is different from an imperial bushel. The distinction was drawn in 1842. (Bushels of grain traded on the commodities market are actually weight measures, and also different.)

Apples and other orchard fruits are often sold in paper bags that hold ¼-peck, ½-peck, 1-peck, and ½-bushel volumes. That's where you'll see these terms most often outside of older recipes.

1 peck = 2 gallons
½ peck = 1 gallon
1 bushel = 4 pecks
1 gallon = 4 quarts
1 quart = 2 pints
1 pint = 2 cups

Tea Towels: I use tea towels like they are going out of style. They cover containers with ferments, cushion hot jars when they come out of the canner, and gently dry off rinsed fruit. I have a bunch, and though they are frayed and stained, they are all clean and ready for battle. Linen or cotton towels absorb liquid better than paper towels, are reusable, and have cute designs.

Canner: A canner is nothing but a big stockpot that will accommodate a canning rack. You can do water bath canning in a regular stockpot, but if you do larger batches, a canner is handy.

Canning Jars: Only buy jars made for canning. Don't can in jars left over from miscellaneous condiments, because they're not made to endure multiple cannings. Save them to store old screws or refrigerate leftovers. One cracked jar can make a giant mess in a water bath canner, and it will waste some of your hard work.

Canning Lids: Most canning lids are two-piece, with a flat lid and a threaded ring that secures it in place. You can reuse the rings, but the National Center for Home Food Preservation recommends only using the lids once.

Jar Lifters: It's possible to extract jars from a pot of boiling water using regular tongs, but it's unsafe and tricky. Jar lifters are cheap. If you do canning, you need these.

Why Buy What You Already Have?

Forage for containers. Your foraged fruit projects—the ferments, the preserves, the dehydrated nibbles—will need homes, and you will be much more likely to make use of them if they land in a logical container.

Mason jars are great for steeping fruit and vinegar or liquor infusions, but they are awful for storing them; I always get sticky dribbles down the sides of the jar when I pour the finished product out. I like to rinse out and save glass vinegar bottles, since they have nice, narrow necks. I also save hot sauce bottles; those flip-top lids with little holes to dole out drops make it a lot easier to get just the amount you want.

If people know you are a container hoarder, they often will happily collect them for you. Recruit people who drink but don't brew kombucha, because glass kombucha bottles are just the right size for bottling shrubs, vinegars, cordials, and sauces. And those giant jars that pickles and fruit salad used to come in are fantastic for ferments or big batches of sangria.

I am snooty, and I think plastic is tacky. But even though I aim to avoid products packaged in plastic tubs, they still amass into towers of disposable effluvia. They are sometimes handy for holding small berry harvests, and I like how the lid keeps berries from getting loose in my backpack. I save everything, though, including bread bags. Why pay for bags when they come free with a loaf of bread? Besides, homemade bread fits into them better than anything else.

Every now and then one of my mom's friends will decide canning is bullshit and they're never going to do it again, and I'll get a windfall of awesome canning jars in hard-to-find patterns. I probably have a few decades before I pass on my treasures to a new generation of container hoarders.

THOUGHTS ON INGREDIENTS

If a chef calls for a specific ingredient, there's probably a good reason. I blithely alter recipes all the time, because that's what makes cooking fun, but before I make tweaks I try to consider what role an ingredient might play, and why there's a certain amount of it. Think of this section as an ingredient decoder.

Sweeteners

Sometimes we want sweeteners with their own character that complements the fruit, and other times ones that aren't noticeable beyond the sweetness they add. If you prefer using natural sweeteners, know that swapping honey or maple syrup won't always give you the same results in texture or taste as refined sugar. But if you do rely primarily on honey or maple syrup for sweetening, you're probably used to the taste and texture differences.

Granulated Sugar: I use granulated sugar for the laserlike precision of its sweetness—that is, it tastes sweet but does not contribute other flavor elements, letting the fruit stand out unadulterated. I used to buy the cheapo store-brand granulated sugar (which often contains less expensive beet sugar), but after comparing that with pure cane sugar, I decided cane sugar has a cleaner taste, so that's what I buy now. Beet sugar and cane sugar do have the same physical properties, though, so either works in a recipe.

Organic Sugar: Organic granulated sugar still has some of its molasses, so it's not pure white. It has a fruitier flavor. Its crystals can be a tad larger than those of conventional granulated sugar and take a little longer to dissolve. You can swap this with conventional granulated sugar.

Brown Sugar: Commercial organic sugar is actually granulated sugar that's had some of its molasses added back to it for flavor. If a recipe in this book calls for dark brown sugar and you only have light (or vice versa), you can swap one for the other without wreaking havoc on the recipe. Like organic granulated sugar, organic brown sugar has slightly larger crystals than its conventional counterpart.

Honey: Honey can be very assertive. I only use it when I want to taste the honey. On top of that, blended and pasteurized honey will taste different from, say, orange blossom honey or tupelo honey. I get raw wildflower honey from a vendor at my local farmers market, and it's always a little different, which is why I like it, but it takes me a little while to get a handle on the personality of a certain batch.

Maple Syrup: Because I adore real maple syrup, I don't use it all the time. It seems a shame to waste it in a recipe where its personality is clouded with tons of other ingredients. But it does often go well with fruit. I like Grade B maple syrup for everything—drizzling and cooking—because the flavor is more robust than Grade A.

Molasses: Molasses is a syrup created during the process of refining sugarcane into granulated sugar. It's bracing and packed with nutrients, though it's still sugar, so it's not a total free pass as far as sugar intake is concerned. It goes well with strong, warming spices like cinnamon, cloves, and nutmeg.

Sorghum Molasses: Sorghum molasses is the juice of a cane, or grass, that's cooked down to a syrup. It's lighter in color than regular molasses, but darker than most honey. It's got a bright, fruity flavor with caramel notes. You can use it in recipes that call for honey or molasses, but gingerbreads won't be as robust with it instead of molasses.

Agave Nectar: I like agave nectar because it's milder than honey, plus it's a vegan alternative. I use it somewhat sparingly, because its floral flavor is sometimes too punchy for me.

Fats

Just as with sweeteners, I try to pair fruits with fats that are a good flavor match. It's pretty amazing to live in an era when it's not unreasonable to have five different types of cooking oil in your pantry.

Butter: Unless otherwise stated, these recipes use unsalted butter. European butter, which has a higher butterfat percentage, will give you different results in baking recipes.

Coconut Oil: Coconut oil tastes like coconuts. I love it, but I try to use it only in recipes where its flavor profile and baking properties make sense. Unrefined coconut oil has a more assertive flavor. Keep that in mind, and use filtered coconut oil if you'd like less of a tropical note. Coconut oil liquefies at a lower temperature than other saturated fats; in the summer mine is always liquid in the jar, and it's easy to pour into a liquid measuring cup and measure. When the weather is cooler, it's solid and difficult to coax out of the jar. Either use a sturdy metal spoon or (if you have a glass jar) set it in a bath of steamy water until it liquefies.

Olive Oil: Mild, fruity-tasting olive oil can be very good in baked fruit desserts—olives themselves are fruit, after all. Olive oils are not created equal. I only use extra-virgin olive oil, because it's less refined. Keep your olive oil in a cool, dark place so it doesn't turn rancid. Good, fresh, unadulterated olive oil can vary a lot in taste, but it shouldn't smell fusty, cardboard-y, or off.

Other Oils: Generally I try to avoid so-called neutral oils like canola oil, grapeseed oil, and vegetable oil—they're highly refined, often extracted with solvents, and usually bleached or deodorized. Having flavorless oil is handy, however, so I do use it from time to time. But I encourage you to explore other oils for cooking and baking: Unrefined peanut oil, avocado oil, walnut oil, and pumpkin seed oil are all fun to experiment with. They taste like the thing they came from, usually.

Other Stuff

Here are a few other useful insights about ingredients that appear a lot in this book.

Flour: All of the recipes in this book calling for all-purpose flour were made with unbleached all-purpose flour, but if bleached flour is all you have it should work fine. Since whole-grain flours can go rancid quickly, it's a good idea to either buy them in small quantities or store them in the freezer or refrigerator.

Salt: It's cool that we have so many varieties of salt available to us now, but it does make things trickier on the recipe development front, since they don't all have the same weight per volume. That said, I stick to plain old industrial table salt in the cheap cardboard canister in my baking recipes. Fine sea salt works, too. When I'm pickling or curing something, I use kosher salt. Once again, the weight per volume of kosher salt varies by brand, but not enough that using a kosher salt different from mine will wreck any of the recipes included in this book.

Cooking Spray: I used to grease baking pans with butter until I read in Rose Levy Beranbaum's *The Cake Bible* that the milk solids in the butter can make baked goods resist releasing from the pan. You are free to grease your pans with whatever substance you like, but I use nonstick cooking spray because it's convenient and does a good job. It's wasteful and barely even food, but a concession I am willing to make.

Vanilla Extract: Vanilla and fruit is my favorite flavor combination. A splash of vanilla extract simultaneously softens fruit's acidic notes and lifts up its bottom notes. I often use it quite generously, up to a tablespoon where some recipes might call for a teaspoon. I know a lot of people who swear by homemade vanilla extract, which they make by steeping vanilla beans in vodka. In my opinion, commercially produced vanilla extract is far superior, with a much more intense flavor, but whatever floats your boat.

Vanilla Beans: Luxurious. These smell and taste quite different from vanilla extract. Since they are not cheap, I use them in applications where their perfume stands out the most, like custards and compotes (the less moisture involved in the recipe, the less the vanilla bean will be apparent). I save vanilla bean hulls that have been steeped in liquid for recipes, then let them dry. When I've amassed enough, I grind them to a powder in a spice grinder. I like to add these ground vanilla beans to fruit butters and compotes. The flavor is muted, but still very different from vanilla extract, and sometimes that muted vanilla bean flavor is exactly what I want.

IMPROVISING

Improvising is often a necessity when working with foraged fruits. What if you have lots of pin cherries and no chokecherries? Red elderberries but not blue or black elderberries? Or only 2 pounds of quince when the recipe calls for 3 pounds? Want to make buffalo berry jelly and can't find a recipe?

I love it when people follow recipes, but I love it even more when they don't. I hardly ever follow recipes to the letter unless I'm developing the recipe or getting paid to test someone else's.

About USDA Hardiness Zones

One note about the "zones" you'll see referred to in part 2 of this book. Hardiness zones are defined geographic areas where certain plants will likely grow because they can tolerate the minimum temperature. The USDA divides the United States and Canada into eleven different zones, each one based on a 10°F (6°C) difference in the average annual minimum temperature. The lower the number of the zone, the lower the temperature a plant must tolerate.

Hardiness zones don't take into account rainfall, light, maximum temperatures, soil conditions, and any other number of factors that make a hospitable environment for a specific plant. Think of the zones as a handy tool, particularly if you'd like to purchase fruit plants for cultivation (most nurseries include hardiness zones in their plant descriptions). But hardiness zones are certainly not written in stone. Our changing climate means the zones are gradually shifting. The USDA updated the zones in 2006, so some older print resources may have incorrect information. If you want to know if a certain plant grows where you live, hardiness zones may be part of your answer, but not all of it. Try to seek out localized information, such as your regional Cooperative Extension System office or state department of forestry, to get the whole story.

If an experiment does not turn out right, then it's on you—but if it turns out great, then wow, what an accomplishment! Be creative. Make pear sauce instead of applesauce, or throw a quince in with an apple pie. Toss a handful of pitted cherries in a baking dish with roasting root vegetables and see what happens.

My big advice on improvising is (a) do it, and (b) do it when you are working in a small scale, so if something is an unequivocal disaster, you won't have wasted a lot of time or expense.

PART TWO

THE FRUITS, AND HOW TO USE THEM

APPLES

Malus spp.
Rosaceae family
Throughout North America in zones 4–8

IT'S LATE AND DARK. You are leaving your favorite bar and as you exit, you notice for the first time that there's an apple tree across the street, and it has dropped apples all over the ground. You scramble around for a few minutes, examining apples under the streetlight for bruises and shoving the good ones into your purse.

Dear reader, I was that after-hours scrumper, at work under the cloak of the evening.

Scrumping is an evocatively dated English term for sneaking apples from an orchard. An article in the *Cheshire Observer* in 1887 described it as "[breaking] (in intent only, of course) the tenth commandment into small bits and fragments." Those guilty of this petty crime generally never lifted prime specimens, and never in large quantities—the word itself is a derivative of *scrump*, meaning "an undersized or shriveled apple."

Perpetrators were likely children playing out of doors, shiftless drifters, and aspiring Romantic poets. Dorothy did it in *The Wizard of Oz*. Could there be a more wholesome image of thievery than teenage Judy Garland in pigtails and a gingham pinafore? The tree she plucked an apple from came alive and smacked her, but no worry: Outside of Oz, apple trees will be happy to release their fruit to you.

There is no fruit better suited to casual five-finger discounts than the apple. One quick snatch and the entire package is secured. It holds up well in a pocket or backpack. Most important, apples grow all over: neighborhoods, farmland, orchards. A resident of a developed suburb is just as likely to come across an apple tree as a person out in the sticks. A small apple is an ideal snack on the fly, and three can be a meal for those who are desperate or easily satisfied.

But scrumping can only get one so far. In late 2007 Linda Bishop, a woman in her early fifties diagnosed as having bipolar disorder with psychosis, wandered for a few days before breaking into a vacant house on an apple orchard after being discharged from New Hampshire Hospital. She lived there for over three months, sustaining herself only on apples. Her cache ran out in December, and she died, weak and starved, less than a month later. The apples had been her savior on the one hand and an agent of her undoing on the other, though ultimately it was the delusions of her illness that led to her demise.

Even in Bishop's self-imposed isolation, the apples were a whiff of connectivity. The trees belonged to someone else; she took what the trees gave and put it to use. The apples were living things linking the two parties. Ideally, that is what apples trees are about: community. The results of this collaboration were quietly tragic in Bishop's case, but in almost all others they are vital and dynamic. Apple trees are social trees.

The global society of apple trees is impossibly diverse. In his seven-volume encyclopedia *The Illustrated History of Apples in the United States and Canada*, apple chronicler Dan Bussey lists seventeen thousand varieties planted here between 1623 and 2000. And that's only one continent. Going beyond that, every apple seed is genetically distinct from its parents. You can't plant a Golden Delicious seed and count on getting Golden Delicious apples from the tree that springs up. Without grafting—inserting the cutting of a chosen tree into the rootstock of another young tree—there would be apple anarchy.

The majority of older American apple varieties were bred for making into other things: cider, baked desserts, applesauce. They needed to withstand months of storage and be naturally resistant to pests—attributes like high tannins or acidity and thick skins were favorable, and looks were secondary. Today appearances are top priority with the comparatively small sector of varieties we see in the grocery store, where our eyes make decisions before our brains do.

Even if a neighborhood tree produces one of the ubiquitous apples in the narrow produce aisle lexicon—Golden Delicious, Gala, Granny Smith—it may look nothing like what you'd get in the store. Community apples are often spotty and knobby on the surface, and they may be smaller or larger than we're accustomed to seeing. Their skins can be mottled with black spots and the pinhole entry/exit marks of worms. These apples need someone to celebrate these attributes. They need you.

The autumn fallen fruit bonanza of apple trees brings us scrumpers out of hiding. It is when our craft takes on an angelic air and becomes gleaning, not stealing. We work alone or in groups, ringing doorbells and calling orchards for permission to collect what's been left behind. We crank cider mills and work our hands sore sliding paring knives between the leathery skin and crisp flesh. We bake and freeze. In Appalachia families get together for the weekend to make gallons and gallons of apple butter, an old but fading tradition. Smaller batches made in slow cookers have eclipsed huge ones in kettles suspended over wood fires, but as in many formidable fruit undertakings, the work of many hands lightens the load and makes it a party.

Apple trees can live a long time, and many a walker of remote areas has stumbled across a ghost orchard—grids of gnarled trees that have endured in their neglect, sometimes producing fruit and sometimes not. A movement of avid cider makers and fruit growers has formed around locating nearly extinct apple varieties, a nitty-gritty effort to rediscover apples that offer flavors our collective tastes have thrown over for something more straightforward. They call themselves fruit explorers, and they keep their eyes out for ghost orchards and backyard trees that give off a mystique. Their obsession with a forgotten past is a grassroots vision of the apple's future.

An apple tree has a magnetic pull. I like to think someone planted it because they hoped other people would find it. Don't fight that pull. Walk the tightrope between scrumping and gleaning, and discover that keeping your balance comes naturally.

Harvesting and Storage

Everything you have just read celebrated picking apples up off the ground. Yes, it's fine. Do it. If you're unsure whether an apple is good, take a bite. If you don't like it, spit it out and move along.

From summer to late fall, enticing varieties of apple trees await you with fruit ready for harvest. Learn which ones to look for in which season (for instance, White Transparents are popular early-harvest apples where I live, while the fabled Gravensteins of Sonoma County are ready in late July).

Scrumped or gleaned apples can rot faster than purchased ones, because they often have bruises and fissures. Refrigerate them once you get home if you don't have the energy to sort through them first. Pick out any damaged ones and use those within a day, lest they go to waste or render that "one bad apple can spoil a barrel" saying to life.

How long apples last in a cool spot, like a root cellar or refrigerator, will depend on the variety. I find it's best to get to them sooner than later, because putting off processing gets easier and easier as the days pass, and suddenly you look at them and realize they've turned to mush.

Culinary Possibilities

Once again, it depends on the apple. Some of the more tannic cider varieties, such as Kingston Black, are not appealing at all eaten out of hand. You get what you get, so taste and go from there.

When all else fails, apple cider and applesauce can save the day. If the apples taste good and are nice and firm, you can dehydrate them; pretreat them with an ascorbic acid solution to keep them from oxidizing.

Fresh Apple Cider

Makes about 1½ quarts (1.5 L)

If you have ever popped popcorn on the stovetop, you know what it feels like to make your own cider. There is not a lot to it, but doing it yourself is thrilling and empowering. You don't need a cider press and an entire orchard—just an electric juicer and some apples. (Fresh apple cider, by the way, is raw, unfiltered apple juice. Hard cider is fermented.)

The owner of a local orchard told me how to do this. "Get a mix of tart apples and sweet apples, then cut them into chunks and run them though a juicer," she said. "I make small batches just like that when we're experimenting with new blends." Green or yellow apples make a pale juice, while red apples yield a russet juice. Gather apples from a few different trees to make your own neighborhood blend.

6–8 pounds (2.5–2.75 kg) apples, preferably a mix of sweet and tart

Wash the apples. Cut out and discard any bruised spots. Halve or quarter them so they'll fit into the juicer's feed tube, but there's no need to core them or worry about the stems.

Run the apples though the juicer. This is a loud, messy, electrified process that will take longer than you think it will. Playing punk rock at a high volume during this step is helpful.

The cider will come out frothy, like a bad latte. It will settle after it sits. Put it in the refrigerator for an hour or so, then skim off and discard any remaining foam. Now clean up the mess, which will be significant. That's the big difference between homemade cider and homemade popcorn.

Keep your cider refrigerated for up to 1 week. You can even freeze it in ice cube trays if you'd like to use a few tablespoons at a time in sauces and the like.

Your fresh, unpasteurized apple cider is the first step to making apple cider vinegar, if you are in the mood to commence a new project. Another storage option is to boil your fresh cider down in a nonreactive saucepan to make a syrupy apple cider reduction, which will keep in the refrigerator for months.

NOTE: You can use a masticating juicer or a centrifugal juicer for this. I'm not sure which will give you the best yield—it depends on your apples and your juicer. If your juicer isn't heavy-duty, I wouldn't risk frying it by running 3 bushels of apples though it.

Life-Changing Applesauce

Makes about 2 quarts (1.9 L)

Homemade applesauce has a velvety texture and complex flavor that blows the store-bought stuff out of the water. It's the perfect destination for gleaned apples, because blemishes don't matter. Better yet, if you have a food mill, there's no need to peel anything. Cooking red apples with their peels on adds a rosy hue.

I've had great success using the mottled Golden Delicious apples from the tree in my friend's backyard—that sauce has an incredible body, and it's just sweet enough that I barely have to add sugar. Make a huge batch of applesauce and can or freeze it, but that might not be necessary with a smaller batch; you'll be surprised how quickly this gets gobbled up.

5 pounds (2.25 kg) apples, rinsed and quartered (peel and core the apples if you won't be using a food mill)

Up to ½ cup granulated sugar or honey, optional

Freshly squeezed lemon juice, optional

Toss the apples into a large, nonreactive pot (such as stainless steel or enameled cast iron) with at least a 5-quart (5 L) capacity. Add enough water to come up the sides of the pot by ½ inch (1.3 cm). Cover and set over medium heat. Once you hear the water boiling, uncover the pot and stir every few minutes, adding a little more water if the apples start to stick to the bottom. Lower the heat, a bit at a time, to maintain a simmer. As the apples cook down, you'll need to stir more often. Some apples will release more liquid than others, so both the cooking time and the amount of water you'll need to add will vary.

Once the apples have totally collapsed (anywhere from 30 minutes to over an hour), pass them in batches though a food mill set over a large bowl (alternatively, push the apples through a colander with a large wooden spoon). Taste the applesauce; if it's too tart, add some sugar or honey in small increments. You may not need any at all. If the sauce tastes flat, add a little lemon juice.

Let the sauce cool before refrigerating or freezing. The applesauce will keep, covered in the refrigerator, for about a week. To can the applesauce, pack the hot sauce into sterilized canning jars, leaving ½ inch (1.3 cm) headspace. Seal and process in a water bath canner for 15 minutes for pints and 20 minutes for quarts.

NOTE: I always make applesauce in a pressure cooker. The end result will be the same, but a few of the steps are different. Put the apples and 1 cup water in the cooker. Lock on the lid, bring to high pressure, and reduce the heat to maintain pressure. Cook for 6 minutes, letting the pressure come down naturally. Carefully remove the lid; the apples should be quite soft, but if they're not, cook at high pressure in 2-minute increments until they are.

Spoon off and reserve the excess cooking liquid, if there is any. Mash with a potato masher or puree or do whatever you like to do to make your apples into applesauce, adding some of the reserved cooking liquid if you need to make it saucier. Taste and adjust the flavor by adding sugar and/or lemon juice.

ALSO TRY WITH: You can also get all mixy-matchy and throw in some pears, Asian pears, crab apples, or quince along with the apples.

Caramel Apple Clafoutis

Serves 8–12

If you come across a tree laden with good, firm baking apples, whip this up. It's a nice change of pace from cinnamon-kissed desserts, and if you are fond of baking things in skillets, it's way faster than tarte tatin.

1½ pounds (680 g) firm apples that hold up well to baking (3–5 apples)
5 tablespoons unsalted butter
⅔ cup (150 g) firmly packed brown sugar
¼ cup (60 ml) water
½ cup (65 g) unbleached all-purpose flour
¼ cup (50 g) granulated sugar
¼ teaspoon salt
4 eggs
1 cup (240 ml) milk
1½ teaspoons vanilla extract

Preheat the oven to 375°F (190°C).

Peel and core the apples. Cut them into wedges about ¾ inch (1.9 cm) thick (you should have about 4 cups/1 L). Set aside.

Melt the butter in a 12-inch (30 cm) ovenproof skillet (preferably cast iron) over medium heat. Add the brown sugar and cook, stirring constantly with a whisk or rubber spatula, until the mixture is shiny and looks like taffy, about 5 minutes. Remove from the heat and stir in the water (the water will spatter—lean away from the skillet when you do this). The sauce may clump up, but it'll smooth out again. Return to the heat, add the apples, and cook, stirring constantly, until the sauce is thick and bubbling, about 5 minutes. Remove from the heat to cool slightly.

Meanwhile, in a medium bowl, whisk together the flour, granulated sugar, and salt. Add the eggs, and whisk until smooth, then whisk in the milk and vanilla. Pour over the apples, and bake until the clafoutis is puffed and set, 25 to 30 minutes. Serve warm with whipped cream or ice cream.

NOTE: If you don't have an ovenproof 12-inch skillet, cook the apples in a large skillet and then transfer them, along with the sauce, to a greased 2-quart (2 L) round or oval baking dish before adding the batter.

Triple Apple Snack Cake with Cider Glaze

Serves 12

Applesauce, apple cider, and fresh apples give this cake a one-two-three apple punch.

For the cake
½ cup (120 ml) applesauce
⅔ cup (160 ml) vegetable oil
½ cup (120 ml) apple cider
¾ cup (165 g) brown sugar
¾ cup (150 g) granulated sugar
2 large eggs
2 medium apples, peeled and diced
    (about 1½–2 cups)
2½ cups (325 g) unbleached
    all-purpose flour
2 teaspoons baking powder
1 teaspoon baking soda
½ teaspoon table salt
¾ teaspoon cinnamon
¼ teaspoon freshly grated nutmeg
¼ teaspoon allspice

For the glaze
1 cup (240 ml) apple cider
3 tablespoons confectioner's sugar

Preheat the oven to 350°F (175°C) and position a rack in the center. Grease a 13 × 9-inch (33 × 23 cm) baking pan.

In a large bowl, whisk the applesauce, vegetable oil, ½ cup cider, brown sugar, granulated sugar, and eggs together until combined. Fold in the diced apples. Set aside.

In a medium bowl, whisk together the flour, baking powder, baking soda, salt, and spices. Add to the bowl with the sugar-cider mixture, and fold until you don't see any dry lumps.

Scrape the batter into the prepared pan, and bake until a wooden skewer inserted in the middle of the cake comes out free of crumbs, 35 to 45 minutes.

While the cake bakes, make the glaze. In a small saucepan, boil 1 cup cider until reduced to ¼ cup (60 ml). Whisk in the confectioner's sugar. Immediately after removing the cake from the oven, pour the glaze over the top and spread it around evenly with spatula. Cool on a wire rack before serving. The cake will keep, tightly covered, for up to 4 days.

MAYAPPLES

Podophyllum peltatum
Berberidaceae family
Eastern US and Canada

Mayapples are prized eating in the foraging world, but I've never had one, and it's not for lack of trying. They are very much a "right place at the right time" kind of fruit, and despite my best efforts, my timing has been off.

A patch of mayapples grew in my parents' backyard—it's still there—right where our lawn met a woodsy area, and I'd play there for hours in the spring, picking mayapple leaves and assembling them into parasols for myself. I called them umbrella pants, and they do resemble toy-sized umbrellas. In all those years, I had no idea those mayapples had little white blossoms in May, just one per plant, and that those flowers would become a fruit no bigger than a walnut.

That explains the name, though in most places the fruit of the mayapples is not ready to harvest in May. A cluster of mayapples is a familiar sight in wildish areas of the eastern United States, but of those many specimens only the plants with Y-shaped stems bear fruits. These are pearly-white oblong things, easy to miss and shyly tucked away under the low canopy of leaves. The fruits are not ripe until they yellow and get somewhat wrinkly, or even fall to the ground, by which time summer is well under way and the pretty green umbrella leaves will have wilted or withered up. To the untrained eye, the ready-to-eat mayapples look like something over the hill and best avoided.

If you find ripe mayapples, don't eat the seeds—they're poisonous in large quantities, as are the stems, leaves, and roots. Native Americans (the plant itself is native to North America) used them medicinally, and they have a laxative effect. Mayapples contain podophyllotoxin, a compound used to make cancer-fighting drugs, as well as a topical drug to treat genital warts.

Despite their playful form, mayapples are not to be consumed lightly; the unripe

fruits are toxic, so lay off if you find a firm one. The fruit is one of those stealthy ones that tastes nothing like you'd expect, a pulpy tropical jelly (or so I hear). While not to everyone's liking, it beguiles some—the intrepid forager's call to arms, basically. I've visited and revisited patches during the spring and summer, and every single time the few unripe fruits I'd spotted weeks earlier had already been snagged by a raccoon or some such critter.

Where you find one mayapple, you'll likely find many, as they colonize by rhizomes in damp, shaded woods. They like border areas between open land and forest, just like the spot where I played as a girl. Mentally bookmark the larger patches you notice in the spring, so you'll remember them when their showy umbrellas shrivel up later in the summer. Mayapple jelly is the classic preparation, though it would take lots of the tiny fruits to cook up a batch. My recommendation: If you find ripe mayapples, nibble them on the spot while reflecting on the overlapping phases of joy they offer: the promise of their green shoots popping up in the early spring, the delight of their fanciful leaves once they reach their full height and unfurl, and the hidden secret of their unusual yet duplicitous fruits once summer hits its full, balmy stride. Also, eschew the seeds.

The Art of Sangria

I love to throw together a batch of sangria with what's on hand. This started when I taught cooking classes. Usually I'd arrive to do prep and find three or four open bottles of wine kicking around from an event the previous night. Then I'd open up the refrigerator and see a bunch of fruit in various states of freshness. Mixing up a big pitcher of sangria freed up space in the fridge and got the students in the class just buzzed enough to be cheerful but not careless.

The key to good sangria (besides a keen balance of tart and sweet flavors) is adding a backbone of quality booze to give the wine some oomph. This could be port, brandy, fruit liqueur, or your own DIY fruit-infused liquor. I like to invest in good booze and then make up the difference with cheap but decent wine. You can get a lot of mileage from one bottle of quality liqueur.

Don't be afraid to mix different wines. That's a lot of what wine makers do—blend wines! Once I saw Jacques Pepin do a Q&A at an event, and a woman asked him what his favorite wine was. "Sometimes, if I have bottles with just a little left, I'll add them together," he said impishly. "Why not?"

You can do the same with your sangria. If I have dibs and dabs of chilled *kompot* or a fruit shrub, I'll toss those in, too. Sangria is a rough formula, not a recipe. Like a great outfit, it's all in how you combine the components.

Harvest Sangria

Serves 12

You know it's a party when there's a giant pitcher of sangria. It's one of my favorite beverages for entertaining because you can throw all kinds of odds and ends in there, as long as they contribute to a harmonious flavor. Good sangria is all about balance. It should be a little puckery, a little boozy, and just sweet enough.

Always make sangria ahead of time, for two reasons: (a) It gives the punch time to mellow, and (b) it's one less thing for you to worry about as you scurry around getting ready for a party. Leftover sangria is awesome.

I like a lot of fruit chunks in my sangria. It's a striking internal garnish, it leaks more and more flavor into the punch as it sits, and when all the punch is gone, you have a boozy fruit salad to pick at later when everyone is kinda drunk. Nourishment!

3 oranges, halved pole to pole and thinly sliced

2 firm but ripe pears or apples, diced

½ cup (120 ml) brandy

4–6 tablespoons granulated sugar, divided

1 cup (240 ml) ruby port

¾ cup (180 ml) apple cider, optional

3 tablespoons lime juice

3 tablespoons lemon juice

2 (750 ml) bottles dry but fruity red wine

In a 1-gallon (4 L) pitcher or glass jar, combine the oranges, pears, brandy, and 4 tablespoons of the sugar. Stir until combined, and let sit for an hour.

Add the port, cider, lime and lemon juice, and wine, and stir to combine, making sure all of the sugar is dissolved. Taste and add more sugar, if needed. Refrigerate for at least 2 hours.

Taste again before serving, adding more sugar or citrus juice, if needed. Leftover sangria will keep, refrigerated, for 3 to 4 days.

ALSO TRY WITH: Plums, peaches, apricots, cherries, loquats, strawberries, blackberries, raspberries, or whatever sweet and luscious fruit is around at the time. The citrus, however, is non-negotiable.

APRICOTS

Prunus armeniaca
Rosaceae family
Throughout North America in zones 5–8

IT IS A BAD IDEA TO EAT apricots around deployed marines who handle armored transport vehicles. According to a 2003 *Wall Street Journal* article, one of the biggest military superstitions among American soldiers stationed in Kuwait near the Iraqi border had to do with a seemingly harmless stone fruit. They were certainly not consumed in any form, and even saying the word aloud could cause a stir among a platoon.

The superstition traces back to a rumor that in World War II, every time a tank broke down, canned apricot C-rations were on board. Marines in Vietnam and Desert Storm have also shared anecdotes about a snack of rogue apricots (always in canned or dried form) preceding a dangerous situation.

But this bit of apricot lore is anomalous. Native to China and Central Asia, apricots are beloved fruits in multiple cultures, and Silk Road traders perhaps first brought the fruit to Persia. Eaten fresh, they are meatier and more intense than any other stone fruit. When dried, they put Clif Bars to shame. When cooked into jam, they render the finest of toast into a mere vehicle for their sticky glory.

Because of the apricot's short season and delicacy, canned or dried versions are what most of us are familiar with. Every Christmas when I was growing up, my mother made *kolachi*, fruit-filled cookies from a recipe given to her by her Polish neighbor. We scoured local grocery stores for Solo brand canned apricot filling, but as the years went on I realized it took less time to make apricot lekvar out of dried apricots myself than it did to go on a wild goose chase for obscure canned ingredients. Not making the cookies was out of the question; using a non-apricot filling even more so.

Spanish missionaries planted apricots in the gardens of California's missions, and today California still leads all US states in apricot production (though the crop pales in comparison with other stone fruit harvests headed to the market). California's apricot crops took off in World War I, when it was not possible to import European apricots.

Any North American fruit connoisseur has doubtlessly heard of the famed Blenheim apricot. Apricots in general have a richness, but

the Blenheim is fabled for its flavor. It was named for the garden at Blenheim Palace, in Oxfordshire, England, where its gardener grew a tree there in the early 1800s. Blenheims are small compared with most other apricots, and not the most profitable tree to farm. Their shoulders maintain a tinge of green when they are fully ripe, and so play tricks on those accustomed to associating ripe fruit with full color break. Blenheims were once well represented in California's apricot orchards, but they eventually gave way to other, better-yielding and longer-lived varieties. Finding fresh apricots on the market is challenging, period, but getting your hands on a fresh Blenheim is a feat.

How dreamy, then, to have one in your own yard or neighborhood! May you enjoy apricots off the tree at least once in your life, and may the experience live up to what others have built you up to hope for. And if you are a marine, may you eat those apricots safely at home.

Harvesting and Storage

Pick apricots when they are ripe. They'll feel softish and should have a nice color (some varieties are more saturated than others when ripe). Apricots on one tree tend to all ripen within the same small window of time. You can pick slightly underripe apricots and get them to ripen up more in a paper bag on the counter, but they won't compare to tree-ripened ones. They'll be a lot easier to transport home initially, though. You need to handle these babies with care.

You can refrigerate ripe apricots for a week or so, allowing you some flexibility with your canning and baking timeline.

To dry apricots, treat them with an ascorbic acid solution first to keep them from turning brown. Firm but ripe ones (not squishy guys) work best. Halve and pit them. You can freeze pitted apricots, but they may brown if not first treated with ascorbic acid.

Culinary Possibilities

Apricots and lamb have what Mediterranean food historian Claudia Roden calls "a special affinity." My cooking school served a lamb tagine with poached dried apricots that I still daydream about. Grilled halved apricots stuffed with goat cheese or crumbled blue cheese are an interesting summer side dish or starter.

Apricots are denser than peaches, and they can make a cloying deep-dish pie. I prefer to bake them in shallow tart pans instead. Almonds and apricots are a power duo; try a frangipane and apricot tart.

Apricot Jam

Makes 4 half-pint jars (960 ml)

No other jam provides as good a reason to get up in the morning. Thick apricot jam is just tart enough to get you going without jolting you too much. There's no need to peel apricots for jam (or for anything, really); their skins are very thin, and they help the flesh hold together just enough to give your final product an appealingly varied texture.

2 pounds (910 g) apricots, pitted and
    roughly chopped
3 cups (600 g) granulated sugar
2 tablespoons lemon juice

Combine the prepped fruit (you should have a little over 4 cups/960 ml) and the sugar in a large, nonreactive pan or Dutch oven. If you like, cover and refrigerate for a few hours, or up to overnight.

Add the lemon juice and set the pan over high heat. Bring to a boil, stirring occasionally, and slowly reduce the heat as the mixture thickens. Some foamy scum may rise to the top during the first 10 minutes or so of cooking. Skim this off with a large metal spoon.

As the jam cooks, you'll need to stir it more and more frequently, and eventually stir it constantly. How long it will take to set will depend on a lot of things: the width of your pan, the heat of your burner, the water content of the fruit, and the volume of your batch (smaller batches of jam take much less time to set).

When it is ready, the jam will be very thick and glossy and register about 220°F (105°C) on a candy thermometer. It will take a little muscle to stir it. Do a gel test on a frozen plate; it shouldn't run, and after a minute or so, if you prod the dollop with your finger, it will wrinkle.

When the jam is set, ladle into sterilized half-pint jars, leaving ¼ inch (6 mm) headspace. Process in a hot-water bath for 10 minutes.

Wild, Wonderful Fruit Preserves

Fruit preserves have their own terminology. Here's how to know what you're getting into.

Preserves, speaking broadly, are foods preserved in jars by canning. Typically preserves are made from fruit—they could be entire kumquats candied in a thick sugar syrup, or peach halves in their own liquid. Usually preserves are whole or chunky.

Jelly is made from the liquid strained out of cooked fruit. Jelly has no solids whatsoever, and it should be clear.

Jam is made from crushed fruit. It may be chunky, and it definitely has solids in it. Jam may or may not have seeds.

Fruit butters are very thick, very sweet fruit purees cooked down until they have a lot of body. Fruit butters are usually made with orchard fruits like apples and pears.

Marmalade is a preserve made with citrus peels and flesh. It is chunky yet spreadable.

Compote is usually not as sweet as jam. It's looser and works well as a sauce. Compote is all about texture. A good compote has a not-too-thick syrup and fruit bits of varying size and color.

Conserve is a preserve made with a variety of fruits. It's a dated term, but a sweet one. An apple and plum jam would qualify as both a jam and a conserve.

Umeboshi-Style Pickled Apricots

Makes enough umeboshi to last a single household
a good, long while

Sometimes a tree dumps fruit on you in manners you can't control. If there's an apricot tree around that's not ripening fruit up to spec, you have an option beyond making the world's biggest crock of apricot butter: Pickle them. *Umeboshi* are Japanese pickles: tiny, shriveled, sour-salty plums. They're made with hard-to-find *ume*, a small Japanese variety of plum, but underripe apricots can stand in for them handily. This idea came from Karen Solomon's book *Asian Pickles*, which you should just get already because it's awesome. While the process outlined here is a little streamlined and not totally authentic, it's still quite rewarding. This is a low-effort project that presents lots of potential. Use umeboshi in dressings, or smear a dab in the center of *onigiri* (rice balls).

2 pounds (910 g) unripe but
    mature apricots
4½ tablespoons (74 g) kosher salt

Rinse off the apricots and put them in a sterile 1-gallon (4 L) nonreactive container (ceramic, wood, or food-grade plastic). Cover them in cool water, and set them in a cool spot to soak overnight.

The following morning, drain off and discard the water. Return the apricots to the container, gently adding the salt in layers so you don't damage the skin of the fruit.

Set a drop weight (such as a pot lid or small plate that's a little less than the diameter of the inside of the container) on top of the apricots, then weigh it down with clean rocks, a large can, or a filled jar. Your weight should be about as heavy as the apricots—for instance, if you have 2 pounds of apricots, your weight should be about 2 pounds. Cover the whole works with a clean kitchen towel, and secure it around the rim with a rubber band. Set in a cool spot and check every day or two. If necessary, gently rearrange the apricots to expose the top ones to the liquid beneath them. If you see spots of white mold growing at any point, remove and discard the mold. The apricots should gradually release enough liquid to be fully immersed. This may take 3 weeks or longer. Once they do, remove the weights.

Replace the kitchen towel, and continue to let the apricots sit in their brine for about 3 weeks. They will slowly become incredibly aromatic, scented gently of

almonds but also of apricots, though not in a sweet way; the smell will be clean, not funky. Taste the brine occasionally; it will be salty, but tart.

Once the apricots are brined to your satisfaction, strain the brine into a clean glass bottle. This is *umesu*, a tart and salty liquid you may use as a seasoning, a cross between a fruit vinegar and a light soy sauce. Refrigerate it for eons.

The umeboshi themselves you can pat dry, put into jars, and refrigerate for at least a year.

NOTE: To determine how much salt to add, calculate 8 percent of the weight of the apricots. You'll add that weight of salt.

ALSO TRY WITH: If you can find ume plums, the small and tart Japanese ones traditionally used for this pickle, go for it! The pickling time may be shorter, since ume are smaller than apricots.

Barley Herb Salad with Apricots and Crispy Halloumi

Serves 8 as a side dish or 4 as a main dish

Pearl barley cooks up plump and chewy—a bit like Israeli couscous, but with more personality. Its subtle sweetness pairs well with fruit. Apricots, whose meaty flesh gives this salad a little brawn, are my preference here, and they play well off the firm and squeaky slices of griddled halloumi, a brined Cypriot cheese that makes the salad a meal (but you can skip the cheese if you like).

½ cup (110 g) pearl barley

1½ cups (360 ml) water

1–2 large ripe tomatoes, diced (about 2 cups)

2 scallions, thinly sliced crosswise

¼ cup (60 ml) freshly squeezed lemon juice

⅓ cup (80 ml) extra-virgin olive oil

1 teaspoon kosher salt

⅛ teaspoon freshly ground black pepper

¼ teaspoon Marash pepper, or ⅛ teaspoon crushed red pepper flakes

⅛ teaspoon cinnamon

4 apricots, ripe but not squishy, cut into bite-sized chunks

½ cucumber, seeded and peeled, diced (about 1 cup)

½ cup (10 g) finely chopped fresh mint

½ cup (10 g) chopped fresh dill

1 cup (50 g) finely chopped fresh parsley leaves

1 (250 g) package halloumi cheese, cut into ¼-inch (6 mm) slices

Put the barley and water in a medium saucepan. Cover, bring to a boil over high heat, and reduce to a low simmer. Cook until the water is absorbed and the grains are plump and fully cooked but still chewy, 25 to 40 minutes. Set aside to cool.

In a large bowl, toss together the tomatoes, scallions, lemon juice, olive oil, salt, black pepper, Marash pepper, and cinnamon. Add the cooled barley, and toss until it's coated with the dressing. Fold in the apricots, cucumber, and herbs. Taste the salad, and add more salt or herbs, if necessary.

To griddle the halloumi, put a 10-inch (25 cm) cast-iron or nonstick skillet over medium-high heat. Once it's hot, add the halloumi slices. Cook them without disturbing too much until they are a nice, mottled brown color (you'll know they're ready to turn when they release easily from the surface of the skillet). Flip them over, reduce the heat to medium, and cook until the other sides are golden brown and crispy, too.

Spoon the salad onto serving dishes, and top each portion with a few slices of halloumi. The salad will keep, covered in the refrigerator, for 4 days.

NOTE: Instead of the pearl barley, you can use 2½ cups of any cooked grain in this salad.

ALSO TRY WITH: Ripe but firm peaches, plums, cherries, or nectarines.

ARONIA/CHOKEBERRIES

Aronia melanocarpa
Rosaceae family
Eastern US and Canada

LOADED WITH ANTIOXIDANTS, the good ol' chokeberry has undergone a superfood rebranding, and now it's known in some circles as aronia. Because chokeberries are so easily misheard (or misread) as choke-cherries, which are a very different fruit, I can get behind this. Aronia sounds pretty, and that's the plant's scientific name, anyway.

Incidentally, why is *chokecherry* a good name and *chokeberry* a bad name? I think it's because *chokecherry* benefits from the charming cha-cha alliteration—*chokecherry*—while *chokeberry* sounds like something you're not supposed to eat. They might as well have been dubbed gagberries or hurlfruit.

Aronia berries may not make you choke, but there's a reason they have remained a specialty fruit. They resemble large blueberries in size, but their skins are thicker and tougher, and their flesh isn't as juicy. And though I've tasted berries much more bitter, only those with galvanized palates rejoice in eating aronia berries raw and unadorned. They are not sweet. They are sour, but their bitterness overwhelms the sourness. Under all of that, they have a promising, dark and wine-y undertone, but you have to think on it a little before it hits you. No wonder some field guides describe these as "technically edible."

This is also why, despite growing over a good portion of North America and even being a minor commercial crop, aronia is not a plant the common person is on a first-name basis with.

A lover of swamps and moist, low woods, aronia is a modest shrub worth looking for in soggy areas. Gardeners and landscapers are catching on, too, and it's gaining popularity as an ornamental. Trusted gardening authority Barbara Damrosch is a fan—kind of. "I will continue to plant these easy-to-grow natives for their fall spectacle, and someday I might even eat one of their fruits—when it's the last berry on Earth," she wrote. All the more berries for aronia converts, I suppose.

Poland, for instance. Poland grows more of the berries commercially than any other country. They've done the majority of the research on the berry's impressive nutritional profile. The European market embraced aronia as a health food long before North American bloggers and natural foodies did. Small farms in the Midwest are also cropping up. Food producers use aronia juice as an additive to wines and juices for color, so there's a chance you've been consuming trace amounts for years.

The color of the mature fruit is the best way to distinguish one species from another. Black aronia ripen earlier, then wither later in the fall and drop from their stems. Red aronia (*A. arbutifolia* / *Photinia pyrifolia*) is found more in the South; its berries may cling to the stems well into the winter.

Harvesting and Storage

Aronia berries are fairly sturdy, and they often grow in decent to abundant proliferations in the wild. You can pile them up in a bucket without fear of crushing. It's okay to be handsy when you pick if speed

The Naming of Fruits

Many fruits that strike us now as quaint or folksy go by various names that are themselves quaint and folksy. That's part of the fun, but it can also be confusing—what's a Juneberry in the South is a serviceberry in the Midwest, and a Saskatoon berry in Canada and the Plains states.

I try to think of these names as part of the plant's story, a story that includes people who came up with the name. So many fruits, particularly more obscure or regional ones, have names that reference another, totally different fruit: mayapples, Hoosier bananas, prickly pears, ground cherries. When we humans name things, we like to do so in a manner that points to something already familiar to us.

Naming is power. In the Bible, Adam names the animals, thereby establishing man's God-granted dominance over them. Sometimes those who wield that power do so carelessly, perhaps not thinking of the repercussions. When we named our daughter Frances, I had no inkling we were sealing her fate to have her name misspelled *Francis* by administrators and teachers her entire life long. If I were allowed to name fruit, they'd all be in honor of members of the Ramones, which might not go over well.

Fruits that have multiple identities can get the short end of the stick, but fortunately some go in and out of fashion. The branding of a large and particularly sunny species of ground cherries as "golden berries" by fruit marketers has greatly increased the fruit's popularity with consumers.

A plant can be branded or trademarked, but that's of no consequence to the plant itself. If one day we are gone but plants are still here on Earth, what will names matter? Maybe every plant has a name for itself, one we can never know.

is of the essence: Grab a cluster and strip it into your bucket. Pick out the stems and leaves later.

Refrigerated, aronia berries keep for over a week. Since they are not excellent snacking berries, consider freezing them; I've heard that thawed berries are less bitter, though I've not noticed this myself. Both thawed and fresh berries can be made into juice. Aronia juice is the starting point for blended drinks, sweetened syrups, fruit preserves, or just having on its own as a tonic.

To make cooked juice, rinse the berries and put them in a nonreactive pot. Add enough water to cover and simmer them until they are soft, about 10 minutes. Mash with a potato masher (or pulse a few times with an immersion blender), and strain through a jelly bag or strainer lined with cheesecloth. (Wear an apron! This stuff is dark and gorgeous and will spatter.) Cool and bottle or freeze.

I like the flavor of aronia better when it's cooked, but it's possible to make raw juice by doing as above, simply skipping the cooking part. The bonus with this method is higher nutrient retention. Give the jelly bag or cheesecloth a good final squeeze to extract as much juice as possible.

Making a straight-up aronia jam or jelly is another option. You could dry the berries, but they wouldn't be very tasty in a trail mix or on cereal, so count on rehydrating them later on for infusions, teas, and such.

Culinary Possibilities

I love aronia berries as character actors in a cast of other fruits. They make things a lot more interesting, but I don't want them center stage. Aronia berries are tart, but lemon juice gives their brooding qualities a flattering jolt of brightness. A drizzle of aronia juice makes a fruit smoothie a deep, princely purple. A modest percentage of aronia berries in a jam of more pedestrian ingredients—say, plums—brings a sneaky complexity and useful pectin, to boot. Aronia juice as a source of pectin in other fruit preserves will make them dark, so keep that in mind, and the amount you'll need to get the jam or jelly to set will require experimentation.

If you are eating aronia berries for their nutritive value but primarily adding them to batters and doughs for quick breads, pancakes, and cookies, perhaps any potential benefits are undone with the dessert surrounding them. Eat the sweets because they are sweet, and stick to aronia juice if you'd rather mainline it.

AUTUMN OLIVES

Elaeagnus umbellata
Elaeagnaceae family
Eastern and central US and Canada, Pacific Northwest

THE AFFECTION YOU DEVELOP toward a forageable has a lot to do with what else is around you at the time. Autumn olives are beloved by some and ignored by many, because there is often something more conventionally delicious around. Flavorwise, autumn olives are not compelling enough to inspire poems and songs and stories—no autumn olive cameos in nursery rhymes, no cul-de-sacs in housing developments called Autumn Olive Place.

Perhaps that's because they are invasive, branded with the Scarlet I. They have not been in North America very long (native to Asia, they were introduced to the United States as an ornamental in 1830), and autumn olive berries themselves are scarlet. Tiny, sweet-tart, juicy, and just a bit astringent, they do not resemble olives in any way except their oblong shape and their single, pitlike seed. They are ripe in the fall—that's the autumn part—but can be ready to pick in the high summer months if it's been hot. At that point they will be juicy but decidedly sour and chalky. Notice their promising fruitiness. The berries stick around on the branch for a while—at least what the birds don't get to—and as time goes on, their appearance doesn't change much, but they become sweeter and mellower. Keep coming back and sampling. Don't give up!

The deciduous shrubs are hearty and quick to grow. Until the 1970s the USDA championed the plant to prevent erosion and create habitat for wildlife. They grow up to 20 feet (6 m) tall, with long branches and leaves with silvery undersides. From a distance those leaves are what give them away. In the spring you'll see small, white trumpet-shaped flowers. The berries are speckled with tiny brown spots, with a matte look, and somewhat pretty in their way. Younger plants can be thorny.

You'll see autumn olive in open meadows and on the edges of woods, or near bodies of water. It has no issues with poor soil, as its roots will increase the nitrogen levels—good news for autumn olive, but bad news for some native plants with different nutrient needs. This is precisely why it is problematic now. Birds eat the berries, poop out the seeds, and spread the plant even more, where it outgrows native plants and blankets them in shade, preventing them from getting the sunlight they need. Their range is from the eastern to the central states, with some showings in the Pacific Northwest.

I suppose if you are eating autumn olives in situ and spitting out the seeds, you are technically part of the problem. The most effective

way to prevent the spread of the plant is to pull up seedlings, or to keep established plants from fruiting (one autumn olive can generate two hundred thousand seeds in a single season). The latter approach requires cutting back as well as periodic application of herbicides.

Autumn olive's relative *E. pungens*, known as thorny olive or silverthorn, is found in the Southeast. It's reputed to be tastier, depending on who the taster is. Those berries are larger than autumn olives. Russian olive (*E. angustifolia*), a tree, is often found in central and western states. Its pale green fruit is also larger than autumn olives, but it is not nearly as tasty.

Some nurseries continue to sell a lot of these plants. Outside of slowing down inevitable shifts in our habitats, a big reason not to introduce them to your property is that it makes eating them less fun. Part of the enjoyment in consuming invasive plants is the illusion of giving 'em what-for—combat eating, a flesh sacrifice to verify our purported dominance over the plant kingdom. That notion of power makes things taste better.

If you don't care for the name *autumn olive*, use their newer moniker, *autumn berries*. After studies discovered that autumn olives are rich in lycopene (the phytonutrient that let ketchup pass as a health food for a while), interest in the fruit's potential as a cash crop picked up a bit, and with that a notion to rebrand.

Harvesting and Storage

Don't pick the berries until they are red—that is, leave the orange and yellow berries on the branch; they are not yet ripe (taste them and you'll know). If a shrub is really heavy with fruit, you can take in quite a haul in no time. Ransack those berries (they're fairly sturdy) by holding an especially loaded branch over a bucket and running your fingers over the fruits to knock them into the bucket. Deal with incidental leaves and debris later. Try laying a tarp under the branches and beating them to detach the fruits (which must be quite ripe for this method to work).

You can even break out the pruning shears and lop off entire branches. Stored on the branch in the refrigerator, they'll keep for longer than you'd expect. In less productive years harvesting is more taxing, requiring a more traditional approach of picking one berry at a time.

The berries freeze well whole. If you have a ton, try drying some and using them as you would dried currants.

Culinary Possibilities

Autumn olive seeds are large and fibrous, but not hard. If the seed is not distracting to you, you can eat the berries whole. Massachusetts chef and author Didi Emmons loves them on her cereal.

Gently cooking the berries and straining the seeds from the pulp gives you some more options. The more you heat them, the more they lose their color. The tannins in the berries cut through rich meats, so try incorporating the puree into a sauce for pâtés or roasted game.

Thanks to all that lycopene, which is not water soluble, autumn olive puree may separate as it sits in the refrigerator. The liquid (Emmons calls it the nectar) will be clear, and the solids will retain most of the color. Even jam may separate once it's canned, though you can always stir it together again when you open the jar. You'll want to use pectin if you do make jam or jelly.

Cooked down gently to a tomato-sauce-like consistency, the strained puree makes a distinctive fruit leather. Or sweeten and freeze the puree in your ice cream maker for a sorbet.

Autumn Olive Fruit Leather

Yield depends

This preparation is a signature of Massachusetts-based wild foods advocate and educator Russ Cohen. He often begins his foraging tours with samples of his autumn olive fruit leather—a smart man! Sharing food is always a great way to begin any social undertaking. This has just one ingredient: autumn olives, and lots of 'em. Dehydrating them concentrates their flavor and their sugar, so they're not as tart as they can be off the branch.

At least 2 quarts (2 L) autumn olives, sorted over and rinsed

To begin with, you'll cook the autumn olives very gently. Put enough water in a large, heavy-bottomed stockpot or Dutch oven to come up about ⅛ inch (3 mm). This will keep the fruit from scorching during its initial cooking. Add the autumn olives, and heat to a slow simmer. Cook for 30 minutes, stirring occasionally, until the fruit softens and the pulp begins to separate from the seed.

Pour the pulp into a food mill or fine-mesh sieve and strain out the seeds. It should have the consistency of thick canned tomato sauce. Line dehydrator trays with fruit leather sheets or parchment paper, and grease them with cooking spray. Pour the strained pulp onto the trays in the thinnest, most even layer possible. Dry at 135°F (55°C) for 12 hours, then rotate the trays (if you're using more than one) and dry another 12 hours, or until pliable and leathery. Cut into sheets and wrap in plastic wrap or parchment paper. The fruit leather will keep as long as you'd like to have it around.

The Invaded Landscape

Aliens, invasives... these are not kind words among many conservationists. Tons of herbicides and thousands of hours of volunteer and paid labor every year contribute to our battle against uninvited plant guests. In many cases plants now considered invasive did not crash the party. We urged them to come here, and then they took over the house.

I have walked past massive curtains of kudzu or expanses of ice plant and given them the stink-eye. I have yanked up garlic mustard and cut down knotweed. I have cursed them and wished them gone so our kinder, nobler native plants could thrive.

In more recent years, though, I've reconsidered this attitude, because the definition of what's invasive and what's not depends a lot on the beholder. "Species come and go so much, as a result of both human and natural forces, that conventional hard distinctions about what belongs where have long been all but meaningless," writes Fred Pearce in *The New Wild*, a book in which he argues that our knee-jerk efforts against invasive species might be misled. Nature, he points out, isn't something that exists as a pristine, perfectly balanced order in perpetuity. Nature is dynamic. Our planet rearranged itself long before we ever did, though we have expedited the pace of the rearranging considerably. Invasive species can be helpful in some cases, and after a period of years the invaded ecosystems may adjust to coexist with their new roommates.

I am trying to see invasive plants through less judgmental eyes, but it does not always work. One thing I am certain of: There is a lot of fruit to be harvested from invasive plants in North America today, and when you come upon it, you can strip the branches bare and not feel one iota of guilt.

BARBERRIES

Berberis spp.
Berberidaceae family
Throughout the US and Canada in zones 3–9

A LESSON ABOUT CASTING JUDGMENT. I have never liked barberry shrubs. We had a Japanese barberry shrub in our yard in Oregon. Not only was it ugly, but with its many sneaky little thorns, it was a bitch to prune, even with gloves. At the time I had no idea of the plant's identity. I just knew I hated it.

Meanwhile, as a serial reader of cookbooks and lavishly photographed cooking magazines, I learned of tiny dried barberries, an important ingredient in Persian cooking, where they are added to pilafs and meat dishes. I always wanted to try some, but I never got to a Middle Eastern market to buy any.

I had no idea that the very shrub in my yard, a Japanese barberry (*B. thunbergii*), was a cousin of *B. vulgaris*, the common (or European) barberry. Though not completely identical, the fruits can be used similarly. The shrub I detested offered something I wanted, after all.

Barberries are a thorny plant in multiple ways, including the plant's invasiveness. Japanese barberry is a popular landscaping shrub because it's hearty and not particularly noticeable. After all, for landscaping to work, you need a mix of starring roles and supporting players. Old-school landscape designers like to use Japanese barberries as extras, particularly in commercial landscaping. A newly erected medical campus I ride my bike past often has dozens of them flanking

their parking lot, bracketed in with mulch dyed a florid shade of red. I never noticed them until I started looking out for barberry plants, because—like extras in a movie—they're not supposed to be noticed.

Birds eat the berries of these non-native ornamentals, spread the seeds all over tarnation, and the adaptable barberry happily establishes itself, crowding out more vulnerable native plants. Some states have outlawed them, along with other introduced species of barberry, including *B. vulgaris*. There are native species, like *B. canadensis*, which you'll find growing in the Midwest and Southeast. I honestly can't be bothered to tell one variety from the other, except that some Japanese barberry shrubs have dark purplish leaves and stems.

In any case, if you live in eastern or central North America, there will likely be plenty of some kind of *Berberis*, quite likely on college campuses, front yards, or the aforementioned medical complex. I also see them growing along my favorite trail in the woods. Are the ones there native or invasive? It's dizzying.

I feel split on barberries, because I truly think they are too ass-ugly to plant on purpose. When trimmed to be symmetrical, they have the patchy look of a clumsily shorn dog. When not pruned, their branches poke out any which way, with unflattering variances in length—a bit like my hair when I first wake up.

Their berries are their saving grace. They are dainty and bright red when color in the natural world is on the wane. After the leaves have turned in the fall and then dropped, ripe, capsule-sized barberries dazzle the eye, tart and teensy. They dangle from the branch by a slender stem. When all of the other, more fun fruits are long gone, barberries will be waiting for you, and they may even get better the longer you wait.

In the spring and part of the summer, the shrubs that fruit will produce berries that start out green and redden as the summer passes. Go ahead and taste one of those red berries in the summer, but brace yourself. It'll be firm and dry with very little flesh and not much flavor to speak of, and what it does have isn't good.

Hidden Realms

There's something irresistible about a fictional world that coexists, unseen by the unsavvy, within the real world we all inhabit. Think of Narnia, or Harry Potter's wizarding world, or the backward land Alice steps into through the looking glass. The completeness of these made-up realms lures us in, particularly when they are infected with something sinister: Jadis, Death Eaters, the screaming Red Queen. Aficionados can read or watch and slip between the two realms—ours and theirs—at will, experiencing the thrill of a narrow escape from Voldemort without suffering from the post-traumatic stress Harry does. Likewise, we can find solace from actual trauma (or boredom) we may face in our ordinary lives.

Consider fruit a *real* hidden world. What fruit is growing out there, waiting for you to find it? Think of the plants you've walked past dozens of times without even taking notice. Once you do notice them, they are a portal to a new realm. They are your looking glass.

The hero of John Carpenter's 1988 movie *They Live* finds a pair of sunglasses that allow him to see the truth—aliens have taken over the planet, and through the lenses he can see their skull-like faces lurking beneath their human disguises. Foraging knowledge and spatial awareness are your own uplifting but equally empowering pair of sunglasses. You put them on and look around and see what's just under the surface of the ordinary. Everyone else might go past a tree or shrub without batting an eye, but you see it as a maker of fruit, giver of life, and part of a larger interconnected system. It certainly beats alien skull faces.

Keep circling back as the season progresses and there's not much foraging to be done. The berries will shrivel up somewhat and remain on that thorny branch. Taste them and check if their character has transformed, with a cranberrylike acidity. Look around. Do you see any other kinds of berries available for foraging? In October or November, I bet not. Might as well give these guys a shot.

Harvesting and Storage

Begin by accepting that you will get pricked, unless you have very fleet fingers. Even gloves, which add a clumsy element, may not be useful. Barberry spines are not particularly large or rigid, but they make up in numbers what they lack in heft. I recall pruning the Japanese barberry shrub in my old yard with gloved hands and getting poked nonetheless.

The color of the berries will depend on the plant; look for either blue or red. Aim for the soft, plump, slightly wrinkled ones. Forager and author Tama Matsuoka Wong says that Japanese barberries are unpalatable in the early fall, but they become sweeter and more complex after the frost sets in.

To harvest a branch laden with berries, you can cut off the branch with pruning shears and then strip the berries off at home. Or strip the berries into a paper bag, leaving the branch attached to the plant. Depending on the shrub, the berries will range in size from small to minute. Go for the biggest ones you can get; they will still be small.

If you'd like to dry barberries, look for the smaller ones, which won't be as seedy. They won't be like the ones in the Middle Eastern market (the seeds in those seem smaller to me, and those barberries, called *zereshk* in Iran, are fruitier), but they could have potential.

Culinary Possibilities

Barberries are not for eating fresh, or for eating in quantity. They are accent berries. Their mealy texture improves with cooking, and with either sugar or a lot of fat. Good, rich plain yogurt dolloped next to meaty braises foils the tartness of the berries, as do onions cooked in tons of butter until brown and limp. Use them fresh as a tart accent in savory dishes—a small handful over roasts during their final stint in the oven will result in burst berries that baste the meat with their juice. The fresh berries can be used like cranberries to make a sauce, but unlike cranberry sauce, it needs to be forced through a food mill to remove the seeds.

BLACKBERRIES

Rubus spp.
Rosaceae family
Throughout the US and Canada

BLACKBERRIES ARE PERHAPS the best known of all foraged wild fruits. Whether they grow modestly on the perimeters of a ramshackle farm or thrive ruthlessly along the banks of a forgotten creek, there are hundreds of hidden wild blackberry havens waiting for opportunistic berry fanatics.

Blackberries exist to lure the weak-willed away from the straight-and-narrow path. Their thorns will scratch, and the company they keep can hurt you. Everything flanking straight-and-narrow paths is bound to be interesting, so I say go for it, but I do have a cautionary tale.

Years ago I had just moved to California and was happily exploring the town of Sonoma, where I had recently set up in a little apartment. I discovered a bike path and trails branching off it into the hills, where my running route wound through madrone groves and next to vineyards. In those days I ran for hours and hours under the blare of the midday summer sun, and it made me a little loopy. When I noticed blackberry brambles not far off the bike path, I got right up in them and gobbled up berries to rehydrate. What I didn't notice in my frenzy was poison oak—blackberries have an affinity for it, as tomatoes do with basil—and the back of my hand must have grazed a cluster right before I used that same hand to wipe giant beads of perspiration off my face.

A week later I had a robust breakout of seeping poison oak blisters on my upper lip. The ooze of the blisters would dry into a crust the color of light amber. It took all of my willpower not to pick at it constantly. While so afflicted, I met Julia Child at a book signing. As she kindly inscribed my copy of *Baking with Julia* and offered earnest advice about a career in food writing, all I could think of was my marred face.

If I had not contracted that poison oak, perhaps I'd have been more receptive to Julia Child's career insights and not floundered around for years working crummy retail jobs and scrounging for oddball freelance gigs. But I'd not be who I am today. And guess what—I still get rashes from overenthusiastically taking off after trailside fruit! I have learned nothing!

The blackberries in question were undoubtedly the invasive Himalayan blackberry brambles that overrun hillsides and choke out native species, but that does not mean the fruit of these dominating opportunists cannot be harvested and eaten with aplomb (just look out for poison oak or poison ivy). Himalayan blackberries (*R. armeniacus*) are not prized for their flavor—I find them sour and wan, though if you're in the middle of running 12 miles (19 km) they hit the spot like nothing else. And if you come across a lot of them, there's always our good friend sugar to make them more palatable.

Disappointingly, the Himalayan blackberry is not from the Himalayas. It originated in Armenia and was introduced to Europe in 1835 for people to cultivate as a crop on purpose, if you can believe that. Like a Gremlin doused with water, it escaped its confinement and rampantly spread throughout the continent. America's own beloved plant maestro Luther Burbank introduced it in America in 1885, likely with no suspicion of how aggressively it would take root all up and down the West Coast. Burbank's aim was to develop fruit and vegetable plants that would withstand long periods of shipping—this was when our nation's transcontinental transportation network was coming into its tween years—so residents of our increasingly urbanized cities could have access to fresh produce. It's Burbank who named it the Himalaya Giant, for the size of the berries. He sold the seeds through his seed catalog.

Feral Himalayan blackberries are deeply intertwined with the cultural identity of modern residents of California and the Pacific Northwest. The thickets are everywhere, at once loved and loathed. Tom Robbins set his 1980 novel *Still Life with Woodpecker* in a Seattle suburb where an exiled king and his family live in a house surrounded with a natural barricade of blackberries. Homeowners and naturalists engage in a never-ending battle with its burly, snaggy tendrils. My

brother, who does non-native plant removal, owes his livelihood in part to Himalayan blackberries.

And yet there are the berries themselves, a seasonal token of redemption for the *R. armeniacus*, a plant impossible to eradicate. Therefore, we must coexist. If the truce lasts only as long as the berries, so be it.

Of course, there are hundreds of varieties of blackberries, native and crossbred. Some have thorns; others don't. Blackberries and raspberries both belong to the genus *Rubus*. Think of them as the patriarch and matriarch of the bramble clan. The extended family of *Rubus* pedigrees (boysenberries, loganberries, tayberries) are considered blackberries regardless of their color, because once picked, they retain their firm white core (or *receptacle*); raspberries don't. This receptacle is why blackberries have a longer shelf life than raspberries—they don't crush as easily.

Late summer is the time for blackberries. Farmed crops start coming into season in the middle of June, but the best wild berries don't start appearing until July, with holdouts ripening into September. An old English folktale warns against picking blackberries after the fall, when the devil makes a mark on their leaves and claims them as his own, although in reality it's more likely birds would have claimed the berries by then anyway.

Gathering fresh blackberries is not without its perils—insects, blazing sun, scratchy weeds—but the rewards are many. Few activities tap so directly into the spirit of summer.

Native to Asia, Europe, and North and South America, blackberries can be found growing on all continents except Antarctica. In Europe and in North America, blackberries have been used for medicinal purposes for hundreds of years; various preparations of blackberry juice, leaves, and bark were said to soothe eye and mouth ailments, aid

digestion, relieve toothaches, and remedy dysentery. Today the focus is more on blackberries' nutritional value: They are rich in antioxidants and dietary fiber.

Harvesting and Storage

Ripe blackberries are deep, dark purple-black—not purple, and certainly not red or green. Berries on a given plant ripen in stages, offering opportunities to revisit a patch to replenish supplies as the weeks pass. Blackberries ripen only on the branch and will not become sweeter during storage. When picked, a ripe blackberry should come free of the plant with nothing more than a gentle nudge. Watch out for thorns, too; not all blackberry bushes have them, but most wild ones do.

Once picked, blackberries don't hold up very long. Blackberries kept at room temperature may mold quickly, so refrigerate them 3 to 4 days, tops; as the blackberries age, they lose their sheen and plumpness, taking on a slightly withered, matte look. Like most other berries, wash them directly prior to eating and no earlier; a premature rinse will lead to mushy berries.

Culinary Possibilities

Barring an all-out bonanza of fresh berry eating, there are two ways to make good on a prodigious blackberry harvest. One is to launch into a frenzy of canning; the other, which is less demanding and more versatile, is to freeze the berries. You don't need pectin to make jam, but many like to add it. Soft or squishy berries that are still good flavorwise are a smart addition to shrubs, sangria, sorbet, compotes, or anything saucy.

LOOK BUT DON'T EAT!

POISON IVY

Toxicodendron radicans
Anacardiaceae family
Eastern and central US and Canada

Not only don't eat, but don't touch. "Leaves of three, let it be," the old saying goes. The berries of poison ivy vines are white, which is usually a dead giveaway for "don't eat me, human!" Many animals, however, make great use of this plant's parts—vines, leaves, and fruits. While you don't want to get too close, poison ivy deserves your attention.

It never even occurred to me that poison ivy had berries until one day, while scrubbing about for elderberries in a brushy area off the side of a busy road, I saw a cluster of small, pearly fruits hanging in a mess of vines that spilled from a giant elm tree to form a tentlike canopy. This leafy tent beckoned me, and I stopped and entered, entranced with the unusual waxy white berries. It was only after recognizing the thick, hairy vines clinging to the elm tree that it dawned on me I was standing under a bonanza of poison ivy that was closing around me like the tentacles of a giant squid. The vines were so mature that the giant leaves didn't seem as recognizable as the young ones I keep my eyes peeled for along the sides of trails.

The blistery rash humans get from poison ivy comes from a chemical in the sap called *urushiol*, but other mammals are immune. Rabbits, deer, mice, and other forest critters eat the young stems and leaves as if they were fancy baby salad greens, and they suffer no breakouts (inside or out). Some birds use the vines as material for nests, bees gather nectar from poison ivy blossoms, and insects eat the leaves.

The tiny berries start out green. In the winter birds feast on the then-ripe white berries. Our bane becomes their savior. Loathing poison ivy and loving wildlife don't have to be mutually exclusive: The plant that has caused me such suffering has also nurtured animals I delight in seeing. There's a children's book about poison ivy called *Leaflets Three, Let It Be!* by Anita Sanchez. In it she writes, "Plants aren't 'bad' or 'good.' They just are!" So it goes with any living thing.

Not everyone is allergic to poison ivy, and it can take repeated exposures before a rash breaks out. Still, you don't want to get cozy. Urushiol is present in all parts of the plant: fruit, leaves, vines, roots. Poison ivy (or poison oak, for you West Coasters) is a reality of foraging, both urban and rural. I often get swept up in the rapture of the moment when I'm foraging, and sometimes I wind up right in the shit with shorts and a tank top, vulnerable as all hell. Because I know I can't count on my good judgment to save the day, I now give my legs (up to the knee) and arms (up to the elbow) a very thorough soapy scrubbing right after I return from a foraging expedition. No fancy soaps or lotions required! This has worked well for me; I've had just a few minor breakouts since.

THE BEST BLACKBERRY PIE IN THE WORLD
Serves 8–12

My mother clipped this recipe from the *Columbus Dispatch* "Reader's Exchange" column decades ago, and to me it is the ne plus ultra of berry pies. I've made it with frozen berries, foraged berries, U-pick berries—you name it. I love how the subtle spices give it a wine-y flavor, and the lemon zest perks up the dark juiciness of the berries.

For this deep and dense pie to set correctly, the filling must come to a rapidly gurgling bubble. When in doubt, err on the side of baking longer.

1 or 2 unbaked Rye Pastry crusts (page 356), or your favorite piecrust recipe

5 cups (725 g) blackberries, thawed if frozen

⅔–1 cup (130–200 g) granulated sugar

¼ cup (35 g) unbleached all-purpose flour, plus more for rolling

½ teaspoon cinnamon

¼ teaspoon nutmeg

Finely grated zest of 1 lemon

Pinch salt

1 egg yolk

1 tablespoon heavy cream

Turbinado or granulated sugar, for sprinkling

ALSO TRY WITH: Raspberries, black raspberries, or mulberries. Also, you can substitute up to 2 cups of blueberries for some of the blackberries.

Preheat the oven to 425°F (220°C). Position racks in the upper and lower thirds of the oven.

Roll out the bottom crust on a floured surface, and lay it in a greased 9-inch (23 cm) deep-dish pie plate. If you're going to make a double-crusted pie, trim the excess crust so it overhangs about an inch (2.5 cm), but don't pleat it. Roll out the top crust, if using, and slide it onto a cookie sheet. Refrigerate both crusts while you prepare the filling.

Taste a few of the berries, and decide how much sugar you'd like to use. If the berries are quite tart, go for the full cup of sugar; if not, add as little as ⅔ cup. Put it in the bowl with the flour, spices, zest, and salt, and stir to combine so there are no lumps of flour. Fold in the berries.

Scrape the filling into the lined plate. Top with the top crust or build a lattice crust on top, and crimp or pleat the edges to seal. Combine the egg yolk and cream in a small bowl, and brush carefully over the top crust. Sprinkle with enough turbinado or granulated sugar so the crust sparkles.

Set the pie on a foil-lined rimmed pizza pan or baking sheet to catch any filling that bubbles over. Set on the bottom rack, and bake 30 minutes. Move to the upper rack, reduce the temperature to 375°F (190°C), and bake 30 to 50 minutes more, until the filling is bubbling vigorously and the crust is prettily browned. (You may need to loosely tent the pie midway through baking to keep the top crust from getting too dark.)

Set the pie on a rack to cool. If you cut into the pie while it is still warm, the filling will be a wonderful gushy mess. It's easier to slice neatly the next day, when you should eat it for breakfast. Serving vanilla ice cream on top is mandatory.

Pie as a Way of Life

Cupcakes, doughnuts, and cookies are only desserts. Pie is a way of life. I don't think there's any other entry in the American pantheon of baking that elicits such devotion and rapture. Pies are good, honest food for good, honest people. They taste best with coffee. You can put ice cream on them and it half melts into a vanilla custard sauce, displaying the appropriate contrast in serving temperatures for the optimal pie experience. Pies are as appropriate at breakfast as they are at midnight.

If a fruit is edible, quite likely there is a pie recipe for it somewhere. The best pies are fruit pies. This cannot be disputed. Nothing showcases the glory of abundant fruit better than a pie: You slice into it and get an eye-catching cross section of gooey fruit glistening like a treasure chest full of gems. The crust, which shears into minute flakes at the gentlest nudge of a fork, is buttery-rich and bone-dry all at once. It is the frame, and the filling is the soul.

My love affair with pies is such that I had to restrain myself from including pie recipes for half the fruits in this book. I prefer berry and orchard fruits in pies (pawpaw and persimmon pies are too custardy for my liking), and a pie composed of nothing but juniper or spicebush berries would be close to impossible, let alone inedible.

The pies in this book are my all-time favorites. I have been making pies long enough now that I improvise both the pastry and the filling, although *improvise* is not the right word. The spiritual warmth of baking projects like pie and raised bread is such that a practiced maker of them intuitively lets her hands do what they need to do; it is almost as if the future incarnation of the pie you are in the process of assembling moves your hands for you.

It would be wonderful to make and eat pie every day, but that is unrealistic for most of us, even those of us who need calories to fuel our daily foraging outings. Pies take time, and they take time away from other things I love. As it stands, I do not make pies for special occasions, but allow the pie itself to be the occasion. That way, if someone asks me how I am, I can simply say, "I ate pie today," and they know that I am well.

Balsamic Blackberry Compote

Makes about 2 cups (480 ml)

Showcase the last berries of summer in a simple spiced compote set off with a drizzle of balsamic. Serve this on rice pudding, panna cotta, or plain yogurt.

3 cups (435 g) blackberries
3 tablespoons granulated sugar
¼ teaspoon cinnamon
Pinch ground nutmeg
½ teaspoon finely grated lemon zest
½ teaspoon balsamic vinegar or
    Cherry Balsamic Vinegar (page 91)

Combine all the ingredients from the berries through the zest in a medium skillet over medium heat. Cook, stirring occasionally, until the berries release their liquid. Simmer 1 to 3 minutes to reduce a little, then crush with a potato masher, leaving half of berries intact. Add the vinegar, and remove from the heat. Serve warm, cold, or at room temperature. Refrigerated, the compote will keep for 1 week.

BLUEBERRIES

Vaccinium spp.
Ericaceae family
Throughout the US and Canada

IF YOU EVER FEEL DOWNTRODDEN by foul weather or a season offering few fruits to collect, take solace in a copy of Robert McCloskey's *Blueberries for Sal*, a classic picture book for children published in 1948. When my own daughter was very small, I became obsessed with it, reading it to her over and over in the soggy winter months that kept us indoors more than I liked. The story offered us a vicarious mother-daughter foraging experience, albeit one with an uncomfortably close encounter with bears.

McCloskey and his young family lived on an island in Maine at the time he created the book, and the story features his actual wife and daughter, Sally—the titular Sal. Sal's mother, looking housewife-comely in penny loafers, a cardigan, and a skirt, sets off to gather wild blueberries on Blueberry Hill. "Now, Sal, you run along and pick your own berries. Mother wants to take her berries home and can them for winter," she tells her daughter, who looks to be between two and three years old. Sal busies herself by gorging on blueberries. (I am glad I didn't have to change that diaper.)

Spoiler alert: Sal wanders off and runs into a mama bear who happens to be eating berries to fatten herself up for the winter. But everything ends up okay, and Sal and her mother cap off the day by canning their foraged berries as planned.

Oh, to have access to fields of wild blueberries! They are a godsend when you encounter them on a mountain hike. No take-along snacks could ever compare. They transform the pace of the proceedings from purposeful trekking to free-form grazing.

Blueberries grow in thirty-six states and are ready for harvest many months of the year at least somewhere. There are highbush blueberries (*V. corymbosum*) and lowbush blueberries (*V. angustifolium*). Lowbush are what Sal and her mom were looking for. As you can guess, they don't grow very tall—a foot or two (30–60 cm), usually—and berries are small and intensely flavored. They grow in more northerly climes (zones 2 through 6) and prefer moist, acid soil but can tolerate dry soil. Maine is the country's biggest producer of wild blueberries, all of them lowbush, and most of them sold in frozen or processed form. They've been harvesting wild blueberries commercially since 1840, in fact. Calling them wild is bending things a bit, because as one farmer told a *Bangor Daily News* reporter in 2010, "We didn't plant them. We manage them."

Commercial farming of highbush blueberries in North America didn't begin until 1912. Prior to that, wild blueberry plants failed to thrive under cultivation. It was the enterprising New Jersey agriculturalist Elizabeth White, daughter of a Pine Barrens farmer who made

important developments in cultivating cranberries, who made the break-through. The younger White took it upon herself to collect samples of the most promising blueberry plants in the Barrens, where they grew heartily in the soil between the bogs. White felt the wild blueberries were too variable to have commercial potential, and she set out to oversee the development of cultivars that would produce sweet, plump blueberries. She partnered with Frederick Vernon Coville, the head botanist for the USDA, and some

of the berries they bred are still widely cultivated today.

Fittingly, the blueberry is New Jersey's state fruit (Maine's state fruit is the *wild* blueberry, an important distinction). The blueberries we buy fresh at the store and pick at U-pick farms are almost definitely highbush blueberries. Highbush shrubs can grow over 12 feet (3.7 m) tall. When my mother took me along to pick blueberries with her at a U-pick farm—I was four and had a couple of years on Sal—I did not eat the berries, but climbed into the tall bushes seeking the biggest, plumpest ones possible. Then I rattled them around in my sawed-off milk jug until they turned to mush. The following year, my mother left me home with Dad.

The ease with which one can buy fresh blueberries year-round today has perhaps diminished the harvesting of wild blueberries in areas where people have proximity to them—and that's the rub. With more developed countryside, their habitat isn't as available as it once was. But keep in mind that cultivated blueberries are bred for many factors, and taste can fall behind the pressing need for productivity, size, and durability. Hybrids that tolerate heat have made it possible to cultivate them on a vast scale in California, Mexico, and South America.

The Pemmican Paradigm

In reference texts about Native American uses for blueberries, pemmican almost always comes up. This low-moisture mixture of pulverized dried lean meat, dried berries, and rendered animal fat was the original energy bar. Nutrient- and calorie-dense and easy to take along on journeys, it kept a long time. Pemmican didn't always include dried berries, but their presence assisted in its keeping qualities and contributed a favorable flavor.

The dried berry component in pemmican varied by location and availability: cranberries, buffalo berries, huckleberries, chokeberries (aronia) and serviceberries were used in pemmican preparations by different tribes. Blueberries in pemmican were special, though, and were generally reserved for celebrations and ceremonies. Today pemmican is experiencing a revival among survivalists and Paleo diet followers. It's a project well suited to modern hunter-gatherers, but it involves a lot of labor, and if you didn't hunt or raise the meat yourself, it is not cheap to make. There is a whiff of privilege—leisure time, disposable income—about its renaissance. Does its growing popularity as a bourgeoisie snack placed in checkout aisles of fancy grocery stores demonstrate aspirations to live a simpler life of eating whole foods, or to get a hot bod that looks good doing Warrior 2 at yoga class?

Pemmican and bark teas. Crushed fruit poultices. Over and over again these uses of native North American plants came up as I researched, but finding particulars was challenging. I wanted specifics, not just fleeting mentions of every imaginable palatable berry being dried and put into pemmican.

Eventually I realized diving deeper would be its own book entirely. The primary sources we have from the time of contact are excerpts from the journals of Europeans who'd arrived here with their own prejudices and notions.

Native foodways were shared orally. The Wampanoag did not enter the menu for the meal cited as the first Thanksgiving into a searchable online database just for my convenience. One aim of the church and government's succession of missions, forced marches, displacement, and reservations was to separate Native Americans from their tribes' traditions, so those are other gaping holes in the chain of foodways. Native peoples were denied access to the foods they had thrived on for generations and instead received low-grade refined flour and rancid oil.

Reducing civilizations to generalizations feels belittling, not only to the past but to the present and future. "When speaking of Native American foodways, it is necessary to take each tribe's regional ecosystem into account, as food procurement and preparation, agricultural methods, and hunting techniques vary widely depending on each tribe's geographic location," writes food historian Ken Albala.

When I find fruit to harvest, I consider that there were methods of preparing it that are lost to the ages. Today, a movement of innovators such as Sean Sherman, a Minneapolis chef who is Oglala Lakota, are combining painstaking historical research with creative guesswork to forge a new Native American cuisine, one that recognizes that while the past cannot be resurrected, it can be respected and reinvented. Ingredients come from the earth. What secrets have they seen with sweeping changes over the ages?

Harvesting and Storage

Blueberries can turn blue a few days before truly ripening; some growers wait until a few berries drop to the ground before descending on the bushes. If you are out in the backcountry, likely you won't be sticking around in that spot very long, so sample what looks good, and it might knock your socks off.

Don't pile blueberries up too high in your container—shallow buckets are fine—but otherwise, they are pretty user-friendly. They'll last in the refrigerator for over a week. Their naturally occurring thin, waxy coating helps give them this longer shelf life. As with almost any berry, don't rinse them until right before you eat or cook with them.

Blueberries freeze well, but they don't keep their shape after thawing. To get dehydrated blueberries with a good texture, you'll want to dip them in boiling water just long enough to crack their skins. Nailing this is trickier than it sounds.

Culinary Possibilities

Probably the best thing to do with blueberries is leave a bowl on the counter and pop them in your mouth periodically throughout the day. Because of the blueberry's year-round availability in grocery stores, few cooks would be hard-pressed to come up with uses for them. Sauces, pies, muffins, smoothies, cereal and yogurt toppers, pancake dotters . . . you know the drill.

Think beyond dessert and sweet breakfast: Pickling blueberries in a light brine is a fun project that gives them a savory edge. A quick blueberry pan sauce can finish seared duck breasts or other rich and fatty meats. Home fermentation enthusiasts can add blueberries to kombucha or kefir.

Roasted Maple Blueberries

Makes about 2¾ cups (660 ml)

Did you ever visit a diner and see a metal caddy on the table housing a variety of pancake syrups? The blueberry syrup was my favorite, though I doubt if the stuff contained either blueberries or maple syrup. This sweet sauce is my answer to that. Roasting the blueberries on a sheet pan reduces their liquid and intensifies their flavor. It's a technique I borrowed from Cheryl Sternman Rule's fantastic book *Yogurt Culture*. And I do like it over yogurt, as well as (of course) pancakes.

3 cups (435 g) fresh blueberries

3 tablespoons maple syrup
   (preferably Grade B)

Preheat the oven to 350°F (175°C).

Line a large, rimmed baking sheet with parchment paper or, better yet, a silicone baking mat. Scatter the blueberries across the sheet in a single layer. Drizzle them with the syrup—no need to toss or stir.

Bake for 15 to 20 minutes, until the berries release their juice. Remove from the oven, let cool for 5 minutes, and then carefully lift the parchment or silicone mat and, curling the ends to make a funnel, empty the contents into a medium bowl or quart-sized jar, using a rubber spatula to scrape all of the good juice off, too. (The berries are easier to transfer when they are still warm.) Cool, cover, and store in the refrigerator for up to 2 weeks.

U-Pick: Not Foraging, but Close Enough

I love U-pick farms. My favorite one is completely ramshackle and totally unorganized. It's on the banks of the Muskingum River, and they grow blackberries, blueberries, and raspberries. I've been to U-pick farms where smooth jazz trios play on Sunday afternoons and they have a small deli with sandwiches for picnicking and it's like a Disneyland of berries. But at this place, you're on your own. The dirt lane to the plants is littered with ancient broken farm equipment. The owners give you vague directions about where to drive to get to which berry, but I always get confused and just wing it. The brambles are as wild and sprawling as Hermione Granger's hair, and the whole thing feels a little bushwhack-y. It's hot when I go there, usually in July, so I wear shorts and a tank top and get all kinds of scratches.

I like to walk in the lanes between the tall brambles. It's like entering a tunnel. At first the picking seems labored, but then it just flows. My brain goes into that special place it goes when I'm in corpse pose at the end of a yoga class. I discover that when my mind is adrift I don't have amazing revelations, but instead all kinds of meaningless flotsam and jetsam rises up with no urgency: snippets of terrible pop songs, interactions from elementary school I'd long forgotten, ideas for essays I'd like to write but never will. Wafts of that special dirty-silty-fertile river smell come up from time to time, and a pontoon or two floats by in the distance.

The farm plays recordings of squawky birds on a loudspeaker to deter real birds from pillaging the berries, and this bird loop cycles through again and again as you pick. I focus on picking raspberries. They grow red, golden, and black ones. I don't like the golden ones as much, but there are fewer of them, and those brambles are patchier. I get this idea fixed in my head that I need to pick an entire quart before I can go home, and before I know it I've spent half an hour trying to fill up my carton. I think I do this because I don't *want* to go home.

I can find blackberries and raspberries out in the true wilds, but never enough for my purposes. It's okay to be a swinger. If you can't pick as much or as often as you like foraging, you can always pick U-pick.

PICKLED BLUEBERRIES

Makes about 4 cups (960 ml)

This simple but addictive pickle comes from chef Joseph Motter and sous chef Sierra Carver of Malabar Farm Restaurant at Ohio's Malabar Farm State Park. Try them on salads, sprinkle on top of a grilled steak, or just snack on a few straight from the fridge; they're surprisingly refreshing.

1 cup (240 ml) red wine vinegar
¼ cup (50 g) granulated sugar
1 teaspoon kosher salt
3 cups (435 g) fresh blueberries

In a 4-cup glass measuring cup or medium nonreactive bowl, whisk together the vinegar, sugar, and salt until dissolved. Add the blueberries to the pickling liquid, cover, and refrigerate for at least 24 hours.

You may either keep the blueberries in the bowl or transfer them to a sterilized quart jar or two pint jars. Refrigerated in their brine, the blueberries will keep up to 4 months. You can also process them in a water bath canner and store them at a cool temperature for up to 2 years.

ALSO TRY WITH: Blackberries, serviceberries, or firm black or red raspberries.

CHERRIES

Prunus spp.
Rosaceae family
Throughout the US and Canada

THE LUSH RED LIPS OF A FAIR MAIDEN are, in song and tale, always compared to cherries. Cheeks are likened to flowers—roses, typically—but lips can never be any other fruit but cherries. Let us consider my favorite shade of lipstick, Revlon's Cherries in the Snow. There's nothing snowy about it, and the hue itself is far too pink-red to approximate the skin of a real cherry, but oh, that name! Such drama!

The idea of cherries pervades our modern culture much more than the reality of cherries, and nowhere is this truer than artificially flavored food. Cherry Popsicles, cherry Dum Dums, cherry slushies, Cherry Jell-O, Cherry Kool-Aid, cherry Jolly Ranchers, cherry Fanta . . . these are the cherry flavors I grew up with and loved. I would not have eaten a real cherry if you paid me in candy. It was only my mother's affection for cherry pie that eventually converted me.

Coming across a tree burdened with thousands of ripe red cherries in the summer is akin to finding a $20 bill on the sidewalk. Nay, a $100 bill! They hang from the tree so brightly, one questions if they really exist even as one stuffs them into one's mouth. Perhaps this idea of cherries as a lipstick color and artificial flavor has distanced us enough that the reality of living cherries is vulgar. And deliciously so.

So many unique cherries, so little space to describe their peculiarities! Because we're more familiar with eating and seeing fresh sweet

cherries such as Bing or Rainier, let's focus on expanding our cherry knowledge with other wild and domesticated ones I hope you'll run into or seek out.

Sour cherries (*P. cerasus*) are beloved of pie bakers and sour beer makers. They originated in Europe and western Asia, just as sweet cherries did. Some sour cherries are not so sour you can't eat them right off the branch. They are juicy, tender, and bracing enough to be interesting but not so sour as to be off-putting. Canned sour cherries make a decent pie, but they can't hold a candle to fresh ones.

My friend's dad has a sour cherry tree in his yard, and he plans his travels so he doesn't miss his annual harvest, which he allocates in zip-top bags for pies throughout the year. He has one of those spiffy cherry-pitting devices where you put a pile of cherries in a hopper and pull a lever, and pop! a pit-less cherry emerges. If you have regular access to many fresh cherries, such a cherry pitter is a worthy investment. Those handheld cherry pitters get the job done, but if you have bushels of fruit, they'll turn your grip to mush.

Compared with sweet cherries, sour cherries are smaller and softer when ripe. Outside of a farmers market, you'd be lucky to encounter fresh ones for sale, which is why sighting a tree is such a score.

If you are not familiar with a specific tree, color is not a good indicator of ripeness with sour cherries, because different varieties can range from vivid red to dark burgundy when ready to pick. They should taste sour (obviously), but not underripe. Montmorency is a popular cultivar that ripens early in the season and has a bright, cheerful color. It's the sour cherry tree of choice for US commercial crops. Most of the US sour cherry crop is grown in Michigan; sour cherries are cold tolerant.

If you can't get your hands on fresh sour cherries, crack open a bottle of kriek lambic. The Belgian-style sour ale is made with sour cherries—Morellos, classically.

Our native North American cherries differ quite a bit from others. They truly do have a wild streak and require more gumption to pursue, but they are fun and funky and worth your time. Black cherries (*P. serotina*) are my favorite, so we'll start with those. Black cherries grow in the eastern and central United States and Canada and can get quite tall (up to 80 feet/24 m). They have a knack for dangling their tiny fruits just out of reach. I see a lot where I live. In the Midwest they often appear where forest meets field and where road meets woods, though maybe that's because I walk in fields and on roads a lot. During the rest of the year, black cherry trees look like regular old trees, but if you look closely you'll see clusters of cherries the size of peas dangling down. A number of cherries will radiate from a central stem. They begin in the early summer as hard green pellets and ripen up right when summer has started to wear out its welcome. The cherries are reminding you to hang in there, because the best is yet to come.

With black cherries you are looking for the plumpest, darkest ones. I find they are sweetest if they're slightly soft. Dry and shriveled ones are too far gone.

If you are out wandering free-form and happen upon a black cherry tree offering ripe, tasty fruit within reach, it's heavenly to collect a handful and then continue on your journey. The best ones will first be sweet, then a bit harsh, and lastly will taste deep and soulful. Suck off the flesh (there won't be much) and spit out the pits in long arcs— they're just the right size to send zooming into the air. The forcefulness of the accompanying *ptoo* sound is very satisfying. Some of those little black cherries won't taste as good as others—they'll make your mouth go all chalky—so just spit those ones right out, flesh and all.

Since the best, ripest black cherries easily drop from their stems, one useful harvesting technique is to spread a sheet or tarp under the branches most laden with nice, ripe cherries and either shake the branch or tap/whack it with a broom handle or long stick (or, better yet, a long stick with a hook at the end, like your teacher used to pull down roll-up maps). Those out-of-reach cherries will drop down and make wonderful percussive taps on the tarp. You'll get a lot of debris, but the time saved in picking is worth it. Go back a few times if the tree is really good, because the fruit does not all ripen at the same rate.

The flesh-to-pit ratio of black cherries does not work in your favor; I figure it's 1:1, if that. These little guys are not for pies, unless you have all the time in the world. These cherries are nicknamed rum cherries because American settlers used them to make wines and infused liqueurs.

Chokecherries (*P. virginiana*) grow from coast to coast. They are easy to misidentify as black cherries, and vice versa. Their range extends farther west than that of black cherries. But since both are edible and, in my opinion, worth looking for, there's no danger in getting the two mixed up. Chokecherries have a wider leaf, tend not to

Looking Up and Looking Down

Not all fruit grows at eye level. Smashed berries on trails and sidewalks can tip you off to overhead treasures to plunder.

Occasionally I'm not able to locate the plant source of dropped fruits, either because I'm in a densely vegetated woods, or because wind played a role in the dispersal. A few times I've excitedly spied apples on the street, only to realize they came not from trees, but humans littering. Generally, though, what came down will lead you to its source. Sometimes when this happens it's too late—the fruit is shriveled and smashed and completely cast off from its branches—but it's certainly not a total loss. Bookmark that spot for next season, and circle back every now and then. Observe what that plant does as it journeys through its seasons.

When fruit prompts you to look up and down, you'll notice other things, too: trees, birds, nifty architectural features that aren't visible at street level. Fruit is just the beginning.

grow as tall, and can be bitterer than black cherries, though not necessarily. They are not as good for eating out of hand, but they are classic cherries for jams and syrups with lots of sugar added. For easy harvesting, chef and midwestern food writer Amy Thielen recommends standing on the bed of a pickup truck, as it's an easily repositioned riser.

Harvesting and Storage

Keep the stems on domesticated cherries. Packed loosely in a single layer, cherries will keep in the refrigerator for about 4 days; day by day, they'll get softer and spottier.

Wild cherries are variable. If black cherries are quite ripe, they will get mushy fast. Refrigerate them when you get home, and deal with them within 24 hours or so. If you want to make jam, you'll need to use pectin.

Culinary Possibilities

Fresh, fried, frozen, preserved, infused in vinegar, baked in pies and cakes . . . cherries are not one of those fruits that present a challenge for those short on creativity. Oh, and often you can just eat them. Do that first.

WILD CHERRY BOUNCE

Makes about 1 quart (960 ml)

Bounce is an old colonial fruit-infused liqueur, usually made with cherries. Spike a batch of sangria with it, add it to cocktails (like a nontraditional Manhattan), or use it in baking. There's no consensus on the base liquor—brandy was popular, as were rum and whiskey—so choose what you like best, and not a top-shelf brand. Decent will do.

This particular formula came from *Charleston Receipts*, a classic community cookbook compiled by the Junior League of Charleston, South Carolina. Its bounce recipe came from one R. Bentham Simons, who stipulates the use of wild cherries and thus immediately won my heart. It's also not as tooth-crackingly sweet as other formulas.

4 cups (560 g) wild cherries such as
    black cherries, pin cherries,
    or chokecherries
1 cup (200 g) granulated sugar, or
    more if needed
4 cups (960 ml) brandy, whiskey, rum,
    bourbon, or vodka

The original recipe says to "Go to the Old Market in June and get a quart of wild cherries." Most likely you will need to gather the wild cherries yourself, and that means they could be luscious and sweet or appealingly sour. Just go with it, and add more sugar if you think you'll need it.

Stem and rinse the cherries. You don't need to pit them, but you may want to mash them with your fingers in a big bowl so the cherries really get in that bounce. I find ripe wild cherries are soft and easily smashed, and the process is messy and fun.

Put the cherries in a large jar with the sugar. Stir them around and let sit for an hour or so, until the sugar dissolves a bit and the mixture is a loose muck. Add the booze of your choice, cover, and let steep in a sunny spot for 10 days. Then move the jar to a dark, cool spot, and let it age a month or so. The color should deepen during this time.

When the bounce has reached the intensity you like, pour it through a cheesecloth-lined strainer and discard the solids, including any sediment. Pour into a sterilized bottle for storage.

ALSO TRY WITH: Plums, peaches, or apricots. Peel the fruit if it's fuzzy or has been sprayed, then chop it. Add the pits if you want some of that almond-y flavor; leave them out for a purer fruit flavor.

NOTE: You can make this with any kind of cultivated cherry instead of wild ones. Presumably cultivated cherries will be larger; instead of smashing them, pierce them a few times with a paring knife. Some people like to snack on the boozy cherries after they've steeped, though I find them a little too mushy for my liking.

Cherry Balsamic Vinegar

Makes about 2 cups (480 ml)

If you're like me and have a black thumb when it comes to making fruit vinegars from scratch, try the easy route and infuse store-bought vinegar with fresh fruit. The slightly woody essence of cherry pits and stems gives cheapo but decent *balsamico industriale* more of a backbone, and the crushed fruit itself perks it up.

3½ cups (395 g) ripe sweet cherries, including the pits and stems
1–1½ cups (240–360 ml) balsamic vinegar

Rinse the cherries, if necessary, and then put them in a deep bowl with a flattish bottom. Mash them gently with a potato masher or your fingers (wear an apron if you don't want red stains on your clothes) until you have a rough, liquidy pulp.

Transfer the cherry pulp, including the pits and stems, to the jar. Add the balsamic vinegar—enough so it tops off the cherry pulp. I like my cherry flavor really strong, so I only use a cup, but for something a little more subtle, add the whole 1½ cups.

Screw on the lid, label and date the jar, and set in a cool, dark place to infuse. Check it after 2 weeks—if you like the flavor, strain out and discard the solids and funnel into a bottle. But probably you'll want it to go longer, so keep on checking it every week until it's to your liking. Three months is totally realistic. When it's ready, strain out and discard the solids though a fine-mesh sieve and bottle the vinegar. Your yield will depend on how long you age the vinegar, how juicy your cherries were, and how much vinegar you added. Store the strained vinegar as you would regular balsamic vinegar: in a cool, dark cupboard for a long, long time.

NOTE: Use cherry balsamic to top fresh fruit, add it to a cherry shrub, dress salads with it, or reduce it on the stove to a syrup and use as a glaze or drizzle.

You can also add a handful of dried cherries in with the crushed fresh cherries to get a sweeter, more concentrated cherry flavor.

Mini Cherry Clafoutis

Makes about 20 mini clafoutis

A cross between a thick crepe and a fruit-dotted custard, clafoutis is hassle-free: no creaming butter and sugar, no sifting flour, and no special baking pans required. In Europe, it's traditional not to pit the cherries, but the pits don't add much flavor, and the dessert is much easier to savor without having to extract cherry pits from your mouth.

I came up with these mini clafoutis (an adaptation of a David Lebovitz recipe) out of necessity; we were making them in a cooking class I was teaching, and I wanted to be sure the dessert had cooled enough to eat before the end of the class. It occurred to me that little ones would bake faster than one large clafoutis. I liked the results enough to do it this way again on purpose. The browned, lacy outer edges of clafoutis are my favorite part, and you get more of them in the muffin tins.

5 cups (565 g) sweet cherries

3 large eggs, at room temperature

½ cup (65 g) unbleached all-purpose flour

1 teaspoon vanilla extract

⅛ teaspoon almond extract

1 tablespoon brandy or kirsch, optional

½ cup (100 g) granulated sugar

1⅓ cups (320 ml) milk

Confectioner's sugar, for dusting

Preheat the oven to 375°F (190°C). Generously grease 20 standard muffin cups.

Stem and pit the cherries. Put four or five cherries in each muffin cup.

Combine the eggs, flour, vanilla and almond extracts, brandy or kirsch (if using), sugar, and milk. Blend until smooth.

Divide the batter among the cups. The tops of the cherries will show, but most of the cherries will be covered in batter. If there are any empty cups, put a few tablespoons of water in each one to ensure even baking and keep the pans from warping.

Bake until the centers are just set and a toothpick comes out free of wet batter, 10 to 15 minutes. The tops will brown somewhat and the centers or sides may puff up, but don't worry; they'll deflate.

Set on racks to cool a bit. If the clafoutis don't pull away from the sides of the tins on their own, take a small metal spatula and slide it around to loosen them. They'll be much easier to remove from the tins at room temperature, and I think these taste best that way, anyhow. Dust with confectioner's sugar, if you like, right before serving.

NOTE: For one large dessert, you may bake the clafoutis in a 2-quart (2 L) shallow baking dish, adjusting the baking time to 45 minutes or so.

Sour Cherry Scones

Makes 10–12 scones

My friend Carrie Havranek of the blog *The Dharma Kitchen* is a scone master, and she came up with a scone to showcase the charms of sour cherries. When these eagerly anticipated little gems show up in mid- to late July, make a batch. Sour cherries are highly perishable and time consuming to pit, but as Carrie says, "Like most things in life that offer challenges, they have their rewards."

2 cups (260 g) unbleached all-purpose flour, plus more for dusting
½ cup (50 g) rolled oats
4 teaspoons baking powder
¼ teaspoon salt
6 tablespoons unsalted butter, frozen and cubed into chunks
½ cup (120 ml) whole milk
Zest and juice of ½ lime
1½ cups (170 g) sour cherries, pitted and halved
⅓ cup (40 g) chopped, roasted, and salted pistachios
Coarse sugar, for sprinkling

Preheat the oven to 400°F (200°C).

In the bowl of a food processor fitted with the metal blade, pulse to combine the flour, oats, baking powder, and salt. Add the frozen butter chunks, and run the processor for 15 to 20 seconds, until the mixture is almost combined; you'll hear some butter bits knocking about the sides of the bowl. Don't overprocess this, or the butter will get too hot and the scones won't hold their shape in the oven.

Transfer to a large bowl, and flour a work surface.

In a glass measuring cup, combine the milk, zest, and juice. Add the cherries and pistachios. Add to the flour-butter mix, and bring everything together gently with a fork to combine. You may need a little more milk, depending on the humidity in your kitchen.

Transfer the dough to a floured work surface, and work it into a shaggy ball. Roll it out into a 10-inch (25 cm) circle. Cut it in half, and then cut each half into triangles about 2 inches (5 cm) wide. You may also just roll out the dough and cut out 3-inch (7.5 cm) circles, if you prefer.

Transfer to a baking sheet lined with parchment or a silicone baking mat and freeze for 15 minutes.

Brush the tops with milk and sprinkle with coarse sugar. Bake until lightly browned at the edges, 12 to 14 minutes.

Remove from the oven, and cool slightly before serving. These are best the day they are made but will keep for a couple of days in an airtight container.

It's the Pits

Have you ever heard that apple seeds are poisonous? More important, why would you want to eat apple seeds? Or for that matter, eat cherry stones or peach pits?

These fruits all belong to the Rosaceae, or rose, family. Their seeds naturally have higher levels of the compound amygdalin, which can in certain circumstances be converted to hydrogen cyanide. Yes, cyanide is poisonous. And yes, some recipes in this book call for cooking apples with their cores, or for infusing cherries and their pits in vinegar or alcohol. What gives?

First of all, a fatal dose of cyanide via fruit consumption would take a lot of fruit. If the seed or pit remains intact, those hazardous compounds stay in the seed. Crushing cherry pits takes a lot of force, so it's unlikely you'll do this in the course of making Cherry Bounce. Apple seeds have a tough protective coating; thoroughly chewing apple seeds would be bad news, but you'd need to consume at least a handful; a stray seed bitten from a core is fine. Likewise, pressing apples for cider wouldn't release enough cyanide to hurt anyone.

As for cooked seeds and pits (which is what happens when, say, you cook whole cling peaches until mushy to make them easier to process into peach butter), hydrogen cyanide is not heat stable and does not survive cooking. Phew!

NOTE: Roasted and salted pistachios impart a savory flavor. If you're using unsalted pistachios, up the salt to ½ teaspoon. If you don't have a food processor, a cheese grater works well for cutting in the butter. You may freeze the shaped, unbaked scones. Bake at 425°F (220°C) until lightly browned; no need to thaw.

ALSO TRY WITH: Blackberries, black raspberries, or dried cherries.

Five-Spice Pickled Cherries

Makes 2 pints (960 ml)

Sweet and tart, these cherries are great for snacking on straight from the fridge. I love the contrast of the warming five-spice with the cool flesh of the cherry. You may can these in a water bath canner, if you like. The brine is well suited to other flavor-forward fruits, too, like plums or even blackberries. This recipe came from Oakland, California–based recipe developer and blogger Leena Trivedi-Grenier. She pits and halves her cherries, while I leave them whole, which means you can't get as many cherries in a jar. If you do as Leena does and pit and halve the cherries, you'll not need as much brine (I relish the pleasure of spitting out the pits, though).

4 cups (455 g) sweet cherries
1¾ cups (420 ml) apple cider vinegar
¾ cup (180 ml) water
1¾ cups (350 g) granulated sugar
1½ teaspoons Chinese five-spice powder

Rinse the cherries. Trim an inch or two (2.5 to 5 cm) from their stems if they are very long (this just makes them easier to pack into the jars). You may also pit and halve the cherries. Pack them into the sterilized jar or jars, leaving ½ inch (1.3 cm) headspace at the top.

Combine the vinegar, water, sugar, and five-spice in a medium saucepan. Bring to a boil over high heat, reduce to a simmer, and cook gently for 20 minutes, until the liquid is somewhat syrupy. Pour the hot syrup over the cherries in the jars.

At this point you may either process the jars in a water bath canner for 15 minutes, or let the jars cool to room temperature before putting on the lids and refrigerating them. Ideally, let the cherries pickle for at least a week before eating them. Refrigerated, the pickled cherries will keep for 1 year.

CRAB APPLES

Malus spp.
Rosaceae family
Throughout the US and Canada in zones 4–8

IF THERE'S ONE CRAB APPLE TREE in your neighborhood, there are a hundred. Decorative and trustworthy, crab apple trees appear in front of houses and schools, or nestled in city parks. They don't grow terribly tall and are ideal for coexisting with humans in developed areas. Sometimes I see them when I'm lost in thickets of trees on properties that have reverted to wild conditions, like mystery trees. Are they truly wild, or were they once in front of a house demolished long ago? Mulling it over is part of the fun.

For an amazing week or two in the spring, crab apples dazzle us with showy white or pink blossoms, and then set about making a fruit that goes generally unloved. There are hundreds of varieties of crab apples, and many of them are useless in culinary terms—tiny, hard as pellets, and unrelentingly tannic.

But a resourceful person who finds a tree bearing crab apples that are larger and juicier will, come fall, never be bored. Hours and hours of picking and goofy kitchen experimentation await.

Crab apples are technically called such because of their size—small—and not their cultivar, since no two apple seeds are genetically alike. Think of them as their own fruit for culinary reasons, since you can't use crab apples exactly as you would larger, more familiar apples.

Unless it is diseased, a crab apple tree will merrily produce crab apples with zero maintenance, though well-timed pruning will keep the trees healthy and shapely. We had a dry summer last year, and while the rest of my favorite off-property fruit trees struggled to produce a fraction of their normal output, there were unlimited crab apples for the taking.

Why, then, are people not beating each other down to get to these things? Because they are a pain in the ass. A harvester can pick ten pecks of apples in the time it takes to collect one peck of crab apples. And obviously you can't just eat them out of hand. Even desirable crab apples are sour. This sourness is presumably the root of their name, given that *crabby* connotes a difficult or grouchy person, though another theory is that it's an alteration of Scots and northern English *scrab*. The term *crab apple* dates from the early fifteenth century.

I am a sucker for underdogs, and so I adore crab apples, but with caveats. Their flesh, when cooked and pureed, is never silky like a good apple applesauce. It takes many crab apples to get much of an end yield, especially if you are coring and chopping their flesh to add to baked goods or chutneys.

Given the seemingly infinite variety of crab apples growing in yards and parks all over, how do you know which ones to target for harvesting? Simple: If you're positive it's a crab apple tree, pick one and bite into it. If you spit it out right away because it's acrid and impossibly tannic, skip that tree. If it's sour but also has actual apple flavor and a crisp, not over-firm flesh, it's got potential. Dark red crab

apples are pretty but usually more tannic.

Always round up the time you expect you'll need to gather any significant amount of crab apples. It can take ages to collect a pound. This is why the residents of the fancy condos downtown see me return again and again to the crab apple trees that grow between their condos and the riverfront bike path. The trees are therapeutically perfect for zoning out under the dappled shade of their leaves at midday, and I like watching people ride by on the bike path. Getting a decent amount of crab apples is part of the point, but not all of the point. If you are lucky a good tree may drop a lot of fruit in good condition all at once, but timing is of the essence in collecting it before it begins to rot.

As for the crab apple trees with fruit too unpalatable to use, I still find great pleasure in them. Many of their fruits cling to the branches all winter long, and some varieties stay red, like tiny cherries, months after the leaves have turned and fallen. They're nature's Christmas ornaments. In some conditions, these persistently present crab apples will slowly shrivel up as if they were air-cured, mellowing their bite and intensifying their sweetness. Take a nibble and if you like the flavor, try using them in boozy infusions.

My favorite pair of crab apple trees grows in front of a house not far from the city's middle school, and the homeowner told me the many kids who pass by after school on their way home make use of the crab apples by pelting each other. Ah, the unbridled tomfoolery of the tween years. Throw them or pick them or simply look. Just don't overlook crab apples.

Harvesting and Storage

As with their full-sized apple cousins, the fruits of different crab apple trees come into maturity from late summer to mid-fall, offering a huge window for crab apple experimentation. Remember: Taste before you pick!

Every time I gather crab apples, mosquitoes come to feast on my bare legs and arms, so dress appropriately or spray yourself with some

noxious chemicals to increase your comfort. Bring along a canvas bag or even a half-peck paper sack to hold the crab apples.

Refrigerate crab apples to extend their longevity, but I find they bruise easily and get spoiled spots within a day or two of storage. Plan to process or cook them not too long after harvesting.

Coring crab apples is tedious as all get-out. I find the best way is to simply cut off all four sides with a paring knife, leaving behind the core. The yield is low, but it's faster than pulling out a doll-sized apple corer.

Culinary Possibilities

Those with little patience for carefully chopping crab apples are best off with preparations that will ultimately be strained or forced through a sieve. Also, some crab apples (just like some apples) turn to mush after just a little simmering, while others hold their shape pretty well.

Think crab apple jelly, or toss a handful in with apples you're cooking down into applesauce. In these cases, the only prep I do is to rinse them off and cut away any spoiled spots. Some recipes say to cut off the stem and blossom ends. I usually don't, but I have found that roughly chopping the crab apples before cooking gives me more juice for my jelly.

Chef Andrew Whitcomb uses crab apples to make *membrillo*, swapping them for quince in the dense fruit paste. Another idea is to juice crab apples as you would for cider, and then ferment the juice for crab apple cider vinegar.

CRAB APPLE CHUTNEY

Makes 2–2½ cups (480–600 ml)

Chopping crab apples is slow going. Here is the only recipe I will do it for. This chutney is great with gamy or smoked meats, or even smoked cheese. It'll also earn a spot at your Thanksgiving spread, whether you serve turkey or not. It's best made with larger crab apples—they are much easier to core—that aren't very tannic. The sage really makes this, but you can use other forceful fresh herbs like thyme or rosemary. If you have fresh sauerkraut around, you can easily make a batch of Curried Crab Apple Sauerkraut Salad (page 104).

4 cups (455 g) crab apples

¼ cup (60 ml) apple cider vinegar

¼ cup (60 ml) honey

¼ cup (55 g) brown sugar

1½ teaspoons whole-grain mustard

¼ cup (29 g) finely chopped
    dried cranberries

1 medium shallot, minced

2 cloves garlic, thinly sliced

1 cinnamon stick

2 cardamom pods

4 dried or fresh spicebush berries,
    finely crushed or chopped, optional

1 tablespoon minced fresh sage,
    thyme, or rosemary leaves

1 teaspoon kosher salt

¼ teaspoon freshly ground
    black pepper

Cut the crab apples from their cores with a paring knife; discard the cores (leave the skins on). Chop the flesh roughly. You should have 2 cups chopped, cored crab apples.

Put the crab apples in a medium nonreactive saucepan. Add the remaining ingredients, and stir to combine. Bring to a boil, cover, reduce the heat to a simmer, and cook 5 minutes. Uncover, and cook 5 to 10 minutes longer, until the crab apples are soft but still hold their shape.

Remove from the heat and let cool (I leave the cinnamon stick in there so it keeps steeping). The chutney is best after its flavors meld for a day or two. It will keep, covered in the refrigerator, for at least a month.

NOTE: The sage is nice if you plan to serve this with poultry, pork, or game (I had this with smoked pheasant breast once, and it blew my mind). The thyme works better for topping cheese and crackers.

If your cranberries are unsweetened, you may want to add a little more honey.

HABANERO CRAB APPLE JELLY

Makes 3–6 half-pint jars (720 ml–1.4 L)

Crab apples, with their high pectin content, make a luxe and silky jelly. The habanero is optional, but its complex fruity and floral character makes for an interesting jelly that's just as well suited to a grown-up PB&J as it is to accompany cream cheese on a toasted bagel. You could substitute minced fresh jalapeño instead.

5 quarts (2.3 kg) fresh crab apples
Up to 4 cups (800 g) granulated sugar
1–4 habanero peppers, stemmed and minced (keep the seeds if you like it very spicy; discard the seeds if you prefer less heat)

Rinse the crab apples well, then sort through them to remove leaves and small branches. Trim off any bruised spots, but leaving the stems and blossom ends on is fine. Halve the crab apples, and put them in a 5- to 6-quart (5 to 6 L) Dutch oven or heavy-bottomed stockpot. Add enough water to cover by an inch (2.5 cm) or so, but not so much that the crab apples are floating all over the place. Bring to a boil. Reduce the heat, and simmer 40 minutes to an hour, until the crab apples are very soft and the liquid is rosy. Remove from the heat.

Transfer to a jelly bag or large colander lined with two layers of cheesecloth set over a large bowl. Strain without disturbing for an hour (don't press out the solids or the final jelly will be cloudy). You should wind up with at least 4 cups of juice. Discard the mushy crab apple solids.

Pour the liquid into the Dutch oven. Bring to a boil, reduce the heat to a simmer, and cook gently for 10 minutes, skimming off any foam with a large metal spoon. Add 1 cup (200 g) sugar for every cup of juice, plus half of the minced peppers, if using. Stir to dissolve. If you have a candy thermometer, clip it on now. Boil gently, periodically skimming off scum as it collects around the rim of the pot. Some of the minced peppers may get skimmed off as you do this, but don't worry; your jelly will still be plenty spicy later.

Cook until the jelly reaches 220°F (105°C) on a candy thermometer or passes the gel test on a chilled plate. This could take up to 45 minutes, so be patient. Add the remaining half of the peppers during the final 10 minutes or so of cooking.

NOTE: Wear gloves when handling habaneros, and avoid touching your eyes and face. Otherwise, you'll have to contend with a burning sensation for the next day or so. Not fun.

Ladle into sterilized jars, leaving ¼ inch (6 mm) headspace, and process in a water bath canner for 10 minutes. Alternatively, you may cool, seal, and store the jelly in the refrigerator for up to 2 months. The heat level of the jelly may mellow as it ages.

Sugar, Foraging, and the Elephant in the Room

"With God, all things are possible." That's the state motto of Ohio, and it comes to mind when I'm researching old recipes for fruit. Here's my version: "With enough refined sugar, all things are palatable." It's the dirty secret of foraging.

Sugar can make food taste good, and it's a staple in my pantry. A default trope of foraged fruits is making some obscure, painfully bitter knob of a berry into jelly. I dig a smear of great jam on toast as much as the next person, but creatively speaking it's the easy way out.

My constantly developing relationship with foraged fruits is a useful reminder of the outsized role refined sugar plays in our modern diets. Granulated sugar is a relatively new player, if you think in terms of human evolution. The whole reason white sugar became accessible to the masses was because of slavery. Life on sugarcane plantations was brutal enough that the mortality rate in the Caribbean kept the slave trade going strong. (In America, where there were not nearly as many sugarcane plantations, slaves lived long enough to reproduce, while on tropical sugarcane plantations slaves were literally worked to death all too often before that could happen.)

I think of sugar's bloody legacy often. Today, its cheap abundance has addicted us to sodas, snack foods, candy, frozen treats, and a bunch of other things I love/hate. Cheap and plentiful sugar is here to stay, at least during our lifetimes, and despite its destructive side, its presence in foods often makes me happy. I love the way molten sugar snaps its glossy bubbles as I cook a pot of fruit preserves. I love how just the right amount of sugar in a bread or cake makes it moist and tender. I love how a strategically tiny amount in a dressing or sauce helps savory tastes assert themselves. The stuff is magic. Take it away, and our entire American way of eating collapses.

To combine granulated sugar—a highly manipulated and industrialized food even if it's organic or raw—with foraged fruits is, in many ways, bizarre. Obviously I still do it, but I try to be mindful of how loaded an ingredient it is, in terms of both its past and its present.

Curried Crab Apple Sauerkraut Salad

Serves 8

The essence of Oktoberfest. Sauerkraut salads are easy to make, and not enjoyed here nearly as often as they should be. You get all the probiotic benefits of eating fresh kraut in a colorful and multidimensional side dish. It's fantastic with grilled sausages, or even small bites on its own. The addition of curry powder is a nod to currywurst, the curry-laced ketchup and grilled sausage dish that's a Berlin street food institution.

1½ cups (360 ml) Crab Apple Chutney (page 101)
¼ cup (55 g) light brown sugar
2 tablespoons vegetable oil
1½ teaspoons curry powder
3 cups (705 g) fresh sauerkraut, rinsed and drained

Mix the chutney, sugar, oil, and curry powder together in a medium bowl. Add the sauerkraut, and mix until thoroughly combined. Transfer to a glass quart jar, and refrigerate at least 24 hours before serving. This will keep in the refrigerator for 1 month.

NOTE: If you make your own sauerkraut and whipping up a batch of crab apple chutney is not your thing, you can throw a handful of stemmed and cored crab apples in a new batch along with the cabbage for some color and a little tartness.

CRANBERRIES

Vaccinium spp.
Ericaceae family
Throughout the US and Canada in zones 3–7*

BOG METAPHORS ARE ALWAYS depressing and menacing. No one likes to be bogged down or have a sinking feeling. Protagonists in fantasy tales journey through bogs as epic character-building slogs to defeat evil or personal demons—think of the Bog of Eternal Stench in the film *Labyrinth,* or Frodo and Sam trudging through the Dead Marshes in *The Two Towers* as they contest with Gollum. Tanned carcasses of Iron Age human sacrifice victims slowly surface in actual bogs, their physical evidence giving us half a story we'll never know the whole of.

If you visit a bog, you will know it is a place of life precisely because it is a place of decay. Organic material breaks down and creates nutrients that allow a diverse cast of plant and animal life to thrive. Bogs are not hospitable to our two human legs, but they are havens for plenty of other species. One of the best loved is the cranberry.

Bogs are acidic wetlands, found mostly in northern US states and Canada, and growing on the sphagnum moss at the right time of year you'll find cranberries like scattered jewels. Generally speaking, there are two types of cranberries: large and small (*V. macrocarpon* and *V. oxycoccos,* respectively). Both are small vines that don't grow upright

* Note that the different species don't all have identical ranges.

and climb the way, say, wild grapes do; rather, they spread out directly over other bog plants.

Sandhill cranes nest in bogs, and surely some were part of the inspiration for one name European colonists had for the cranberry: the crane berry. Tiny white and red cranberry flowers have narrow petals that swoop upward, suggesting the form of a crane.

Lowbush cranberries (*Vaccinium vitis-idaea*) are small evergreen shrubs native to Europe and North America. They favor tundra and boreal forest. You may know the lowbush cranberry better as lingonberry, mountain cranberry, or red whortleberry. They are a favorite in Scandinavian cuisine, as anyone who has spent time in IKEA's food marketplace knows.

The cranberries we know best as consumers are cultivars of *V. macrocarpon*. Cranberries were one of the earliest crops grown commercially in North America. In 1816 Massachusetts farmer Henry Hall noticed that sand blown from the nearby coast over his cranberry vines helped them thrive, so he created beds to mimic this phenomenon and had great success.

Most of the cranberries grown in America today are destined for bottled "cranberry juice cocktail," which is heavily sweetened. If you have ever suffered through a urinary tract infection, it's possible you rushed to the store to buy a bottle. Cranberries have long been used to prevent this malady, beginning with Mohawk and Inuit tribes. But cranberry intake alone cannot stop a current infection, alas; it works by keeping future ones from happening, and supporting the healing during recovery.

Thanksgiving is the holiday when we demand foods made with fresh cranberries. There's the quivery cylinder of polarizing canned sauce, yes, but also cooked and raw relishes. Just like they had at the first Thanksgiving, right? Well, no. The Wampanoag did introduce the Puritans to cranberries, but they were not served at the meal we now call the first Thanksgiving. Food historian Sandy Oliver suspects that Native Americans didn't combine cranberries with maple syrup or honey "in the kinds of culinary applications we are familiar with." She feels it's more likely they used them in stews and meat preparations.

Cranberries float because of the air pockets inside them. You may have noticed these if you've ever chopped fresh cranberries; they have a different cross-section from other *Vaccinium* species. This buoyancy helps in harvesting—growers flood the bogs so the berries float to the top, where mechanical pickers comb them off the vines like lawn mowers. Earlier harvesters used handheld cranberry rakes that resemble the bucket of an excavator. You can run into wooden ones at antiques shops, and order new ones (usually metal or plastic and metal) to use in your own harvesting.

Those watery red fields of cranberries undulating over flooded beds are a striking image in commercials on television, but they are nothing like what it is like to harvest cranberries in the wild. In the fall cranberries are your lure to go check out the gushy, living expanses of bogs, where you can experience a vital, hopeful world far removed from our own.

Harvesting and Storage

Don't wear your nice clothes to go cranberry picking. You will be doing a lot of bending over. Cranberry rakes are meant for use in flooded bogs, and you won't be flooding any wild bogs, so picking the old-fashioned way is how it's gonna be unless the cranberries are growing somewhat erect.

The fruits are hardy, and if they freeze on the vine, you can go get them after a thaw and use them as you would commercially frozen cranberries; that is, they won't be firm like when they were fresh, but they will be fine for cooking or baking.

Cranberries will keep in the refrigerator for weeks, and in the freezer for up to a year. Don't rinse them off before freezing them; just dump them in a bag, seal it, and pop it right in there.

Drying cranberries is tricky. A quick blanching will break their skins and allow them to dry more evenly in a dehydrator.

Culinary Possibilities

Sweetened dried cranberries have their charm, and cranberry juice is a hero of punches and cocktails, but it's the fresh berry that offers the most potential. Their bite makes savory-sweet sauces ideal for cutting through rich holiday foods, and their color enlivens the palette of fall feasts. Throw a handful of fresh cranberries in a pan of Thanksgiving dressing (that's what it's called when it's not in the bird) before it goes in the oven.

Don't forsake cranberries once there's no longer any leftover sauce from Thanksgiving kicking around. I make a fermented raw cranberry relish that I plop on bowls of grains and vegetables all winter long. It provides just the jolt a person needs when the gray days of fall set in and the trees outside look like skeletons.

Cranberries are high in pectin. That's why cranberry sauces—both cooked and uncooked—set so well and have that gelatinlike

quality. Tarts and pies with deep-dish cooked cranberry fillings will have an appealing, jammy texture because of this. Often they are cut with apples, pears, or other sweeter fruits to curb the cranberry's potential to overpower.

Fermented Cranberry Relish

Makes about 2½ cups (600 ml)

This is very similar to the classic raw cranberry relish, but given a boost from a stint of lacto-fermentation. The fermentation imparts a better texture and adds a little life (literally) to the typical November feast. It's now my cranberry relish go-to, a salvation for dry leftover turkey. But don't stop there—in the post-Thanksgiving weeks I spoon big dabs of it over all kinds of crazy stuff, like sauerkraut soup, black-eyed peas with rice, and (the best) classic turkey or chicken soup. It really brightens up food, in terms of both flavor and color.

A food grinder is the ideal tool for making this relish, but a food processor will work just fine. After fermentation, if the relish is too tart, add more sugar; if it's too sweet, stir in a little lime juice.

1 large orange
2 cups (200 g) fresh cranberries
1 medium jalapeño or habanero pepper, cored (for a milder sauce, remove the seeds and membranes)
⅓ cup (65 g) granulated sugar
½ teaspoon kosher salt
1 teaspoon yogurt whey, optional

NOTE: Wear gloves when handling habaneros, and refrain from touching your eyes and face to avoid irritation. You can skip the fermentation, if you like, but the relish is still better a few days after you make it.

GRINDER METHOD: Roughly chop the orange. Combine with the cranberries and pepper, and run through the coarse or medium die of a food grinder. Mix in the sugar and salt. If you have yogurt whey, you can add it to help jump-start the ferment.

FOOD PROCESSOR METHOD: With a sharp knife, cut the top and bottom off the orange. Stand the orange up on its cut end and cut away the remaining peel. Mince the peel by hand, and roughly chop the flesh. Place the cranberries in the bowl of a food processor, and pulse a few times. Add remaining ingredients (including the orange flesh and peel), and pulse a few more times until the mixture is well combined but not pureed.

Whichever method you've used, once you have your relish, pack it into one or two clean glass jars and screw on the lid(s) until they are fingertip tight. Leave out at room temperature for at least 1 day and up to 3 days, checking periodically. You want to see some bubbles and taste a sprightly flavor. The best way to figure out if it's fermenting is to hold the open jar to your ear, like a seashell; you should hear faint fizzing. Store in the refrigerator for up to 1 month. (I've kept this for up to 6 months, but the flavor gets less punchy as time passes.)

Cranberry Ketchup

Makes about 2 half-pint jars (480 ml)

This tangy condiment is based on a recipe in Time-Life Books' *American Cooking: The Northwest*. Though it shares seasonings such as cloves and allspice with its tomato-based cousin, cranberry ketchup's texture is firmer and more gelatinous, with a pleasingly puckery finish. Serve this with wild game, rustic pâtés, and terrines. For a snack to reckon with, top a thin slice of pumpernickel with sharp aged cheddar and generously smear with cranberry ketchup.

6 cups (600 g) fresh cranberries
1 large onion, chopped
1 (10-inch/25 cm) rib celery,
    including leaves, chopped
1 cup (240 ml) water
⅔ cup (160 ml) apple cider vinegar
1 cup (240 ml) honey
½ cup (100 g) granulated sugar
¾ teaspoon ground cloves
1 teaspoon ground allspice
1 teaspoon cinnamon
1 teaspoon salt

Rinse the cranberries with cold water (if they're frozen, no need for you to thaw). Place them in a large stainless-steel or enameled saucepan with the onion, celery, and water. Bring to a boil over medium-high heat. Cover, and cook for 10 to 12 minutes, until the cranberries are very easily mashed against the side of the pan with a spoon.

Run the cranberry mixture through the fine blade of a food mill or rub it through a fine sieve with the back of a spoon, pressing hard to release as much puree as possible. Discard the solids.

Return the puree to the pan, and add the remaining ingredients. Cook over medium heat, stirring frequently and skimming off any scum, until it's thickened enough to hold its shape on the tip of the spoon, about 15 minutes. Taste to check for seasonings. Ladle into two sterilized half-pint glass jars, leaving ¼ inch (6 mm) headspace. If you'd like to can the ketchup, seal and process in a water bath canner for 10 minutes. If you're not canning, cool, cover, and refrigerate for up to 2 months. Allow the ketchup to mellow in the jar a few days before opening.

HIGHBUSH CRANBERRY

Viburnum trilobum
Caprifoliaceae family
Northern US and Canada

Highbush cranberries are not actually cranberries at all, but cranberry poseurs. They're in the honeysuckle family, and bog cranberries are in the heath family. The fruits of both plants have similar-ish appearances and flavors, and they're both native to North America. Hearty and deciduous, highbush cranberries can grow up to 15 feet (4.6 m) tall.

They're planted as ornamentals and like moist but well-drained sites, but in the wild they can be found in damp woods, along streams, and on woodsy slopes. The berries ripen earlier than true cranberries and can be used similarly, since they are acidic as well. They do have a large seed, and cooked fruits will benefit from straining.

CRANBERRY STREUSEL BARS

Makes 36 bars

Anything with streusel is messy to eat, but that crumbly and sweet topping can make a dessert. It contrasts nicely with the jammy and tart filling. My mom makes a thematically similar recipe with dates and orange zest in the cranberry filling, but I am a purist, and I like to let the cranberries do the talking here unaided.

Cranberry Filling
3½ cups (350 g) fresh cranberries
¾ cup (180 ml) water
1 cup (200 g) granulated sugar
Pinch salt
¼ teaspoon ground mace

Crust and Streusel
1 cup (220 g) brown sugar
1¾ cups (230 g) all-purpose flour
½ teaspoon baking soda
½ teaspoon salt
½ teaspoon cinnamon
¾ cup (1½ sticks) unsalted butter,
    at room temperature
1½ cups (150 g) old-fashioned oats
    (not instant)
½ cup (60 g) chopped walnuts or
    pecans, optional

Preheat the oven to 375°F (190°C). Grease a 9 × 13-inch (23 × 33 cm) baking pan and set aside.

To make the filling, combine the cranberries, water, and sugar in a medium saucepan. Bring to a boil over high heat, then reduce to a simmer. Cook, stirring occasionally, until the mixture thickens and the cranberries pop, about 5 minutes. Remove from the heat, and stir in the salt and mace. (The mixture will thicken as it cools.) Set aside while you make the streusel.

In a large bowl, combine the sugar, flour, baking soda, salt, and cinnamon. Add the butter and, using your fingers, work together until no small lumps remain. Add the oats.

Press and flatten half of the streusel mixture into the pan to form a crust. Bake for 10 minutes.

Remove the pan from the oven and spread all of the cranberry filling over the hot crust. Add the nuts to the remaining streusel, if you're using them; sprinkle the streusel evenly over filling. Return to the oven, and bake 25 minutes, or until the streusel topping is lightly browned. Set on a wire rack to cool completely, then cut into bars. Store in a covered container for up to a week, or freeze for up to a month.

NOTE: If you have about 3 cups (720 ml) of leftover classic homemade cranberry sauce—the cooked kind that's sweet and doesn't have bits of celery or onions—just use that for the filling.

Why Pectin Matters

Pectin is a soluble fiber found in plants. It's what gives fruit preserves their gelatinous body. Not all fruits have the same levels of pectin, so if you want to make preserves, it's important to understand how pectin works, and where to find it.

Underripe or sour fruits (such as crab apples, quince, lemons, and cranberries) have the most pectin, and fruit in various stages of ripeness will have varying levels of pectin. Green gooseberries have more pectin than the riper reddish ones, for example. Jam makers can use this to their advantage and deliberately include both. This is why it's a good idea to throw some strawberries with hard, white tips in with those nice and soft and sweet ones when you're making a batch of jam.

Some fruits, like nectarines, have low naturally occurring pectin content and need either added pectin or added acid to set. Pectin, sugar, and acid all act on one another in fruit preserves to make a gel. Adding sugar, which is hydroscopic, removes enough water to encourage molecules to bond together, and acid (usually in the form of lemon juice) further helps the pectin molecules gel.

Getting just the right combination of the three—pectin, sugar, and acid—is the tricky magic of making preserves. A lot of preserve makers almost always add isolated pectin in some form as an insurance against a runny batch. Commercial pectin comes in powered or liquid form. It's usually extracted from citrus or apples. It can't do its special juju if there's not enough sugar in the preserves, so

there are low-sugar "lite" pectins for such situations. If a recipe calls for one type of pectin, that's what it was formulated for, so if you only have powdered and the ingredients call for liquid, it's not advisable to swap.

I'm honestly a terrible person to be telling you this, because I don't use commercial pectin at all. I have no beef with it, but I tend to be drawn to making preserves with high-pectin fruits in the first place. I also like the challenge of getting the preserves to gel with other methods. You can make your own liquid pectin from apples or crab apples. Citrus seeds gathered into a cheesecloth bundle can help a marmalade set. My favorite trick is to mix high-pectin and low-pectin fruits in a batch together. Aronia berries (chokeberries) are high in pectin and add a nice color to pale fruits, so I might toss a handful in my kettle with the rest of the mess when I get started.

CURRANTS

Ribes spp.
Grossulariaceae family
Throughout the US and Canada

CURRANTS ARE LIKE TINY GEMS, and seemingly about as rare. Good luck finding any at a market, because they have a fleeting shelf life and are, in some states, illegal. Also, they are too tart to the tastes of many to be enjoyed fresh. Rare? Tart? Illegal? People whose ears prick up when issued such challenges do, or will, love currants.

Unlike their kissing cousins, gooseberries, currants are hairless. Their leaves and twigs may have a skunky smell when you rub up against them or prune them. They don't mind shade, unlike a lot of other well-loved berry plants. They like coolish summers and cold winters. The fruit itself grows in long, dangling clusters, like ritzy drop earrings. For your flavor and fashion delight, they come in red, black, white, and golden styles.

We have native currants and domesticated European varieties growing in North America, but on different scales. Anyone with a glancing knowledge of currants (a small segment of the population, granted) will bring up the whole illegal thing pretty quickly, as it puts currants in a suspect light.

Around 1900 a fungus called white pine blister rust from imported French pine seedlings began striking forests in the Northeast, and a government-sponsored eradication program began. *Ribes* like gooseberries and currants are also affected by and carry white pine blister rust.

The fungus *Cronartium ribicola* needs two plants to complete its life cycle: a five-needled pine and a plant belonging to *Ribes*. The fungus does more harm to the pine tree—it can be fatal. The *Ribes* come out comparatively unscathed. Black currants were singled out as a particular threat. Beginning in the 1920s the federal government issued a blanket ban against cultivating or importing all *Ribes*. Like the prohibition of alcohol, it altered the landscape of cultivated fruit in America.

In 1966 the federal government lifted the US ban, but individual states such as Delaware, Maine, Massachusetts, and New Jersey still have limitations on either black currants or even all *Ribes*. Plant breeders in Russia developed currants that are resistant to white pine blister rust, and the home gardener in pro-*Ribes* states can join in the slow reintroduction of currants to the North American diet. Your state extension service is the best resource for learning if there are any prohibitions where you live.

The fungus threatened the white pines of Europe, too, but since white pine was not as important there economically, residents placed black currants before pines. Europe today produces over 99 percent of the world's currant crop. More than half of it goes to juice. When was the last time you cracked open a refreshing black currant punch? You might have to visit a nifty import shop to do it. Ribena is a sugary bottled black currant beverage well loved in England and once touted as a healthy way to pump little ones full of vitamin C. Maybe it's like our Hawaiian Punch. Some drink unadulterated black currant juice or extract as a health tonic for its hearty antioxidant content.

Our own native black currants grow in eastern and central Canada and the United States, as far west as the Rocky Mountains and as far south as Virginia. Black currants are somewhat elderberrylike in that cooking changes them from a nose-scrunching disappointment to a richly nuanced delight. Simply simmering black currants until they break apart and pressing the works through a sieve yields the base for a syruplike coulis to sweeten and drizzle over custards or ice cream.

Red currants are sweeter. Once again, there are both native and European varieties. White currants, which are "as rare as truffles," as food writer Marion Gorman put it, are just a strain of red currants. If you get

HONEYSUCKLE

Lonicera spp.
Caprifoliaceae family
Throughout the US and Canada

An overenthusiastic and incautious forager can mistake this trickster for a number of edible berries. That's right—honeysuckle, beloved for its fragrant blossoms and honey-tinged nectar, produces mildly poisonous berries.

Honeysuckle berries range in color from yellow to black, depending on the ripeness and variety. The ones I see most are red and could be mistaken for currants or autumn olives by those who are ill informed or not wearing their corrective eyewear. The berries grow directly from the stem in little clusters and can be paired like conjoined twins.

Honeysuckle is a non-native plant considered invasive in some areas. It was introduced as an ornamental and also for ground cover and weed control. Now, depending on where it's growing, it's unwelcome itself.

There's a romance about honeysuckle that makes it easy to love, though. My parents still have it growing up a trellis in their backyard, and I recall picking the blossoms and pinching off the base to squeeze out a few drops of that mind-blowing nectar. Good news: The nectar is *not* poisonous.

your hands on fresh ones, whether red, golden, or white, drape clusters of them over plated desserts as a garnish; their simple beauty elevates a dish.

The dark, dried currants most of us know resemble little raisins, because they are, in fact, raisins; that is, they are dried Zante grapes and not dried currants. This does not mean you can't try real currants.

Harvesting and Storage

Currants won't go bad if they're left on the plant for a little while, which is to your advantage. The longer they're left on the plant, the sweeter they get, although of course they will eventually shrivel up and rot. The USDA Plant Guide says that currants are harvested by "flailing them into containers as soon as they ripen," which sounds pretty badass, so maybe try that, too.

Since the colors of currants vary, taste first to determine if they're ready. If you feel they are ripe, you can cut off entire stems of clusters (if you will be cooking the currants, you can even leave those stems on).

Once harvested, deal with those currants sooner than later. Refrigeration will give you some wiggle room, but these beauties don't like to linger. Process them or freeze them, though after thawing their shape won't hold as well.

Culinary Possibilities

Because of their seediness, currants meet their end in jellies far more often than in jams. Red currant jelly is *the* classic base for glazing those colorful, glistening French-style fresh fruit tarts. You can add it to braised red cabbage for the ultimate side for sauerbraten.

Enjoy red currants raw, or crush them with other berries and pack them into a mold lined with thinly sliced bread for a classic English summer pudding. The bread absorbs the berry juice during an overnight rest, and the following day you liberate a dome that gushes when you cut it. It's way better than I'm making it sound, and utterly edible on a hot summer day.

Black currants are usually best cooked. Cordials, preserves, or compotes to serve with desserts or game meats and charcuterie are all solid options.

ELDERBERRIES

Sambucus spp.
Adoxaceae family
Throughout the US and Canada

ELDERBERRY SHRUBS OFFER both yin and yang, producing two beloved and distinct ingredients. Elder*flowers* have a flavor as delicate and dainty as their appearance; elder*berries* are powerfully purple and, when paired generously with sugar, taste rich and forceful. Harvesting the flowers will diminish the potential crop of berries you'll get later in the summer, but if you are lucky you'll spot enough elderberry shrubs or trees that supply won't be an issue.

Supply is an issue for me; elderberries do grow wild on my favorite stomping grounds, but not in numbers sufficient to make use of them. I enjoy seeing the flowers in the spring and monitoring the progress of the tiny berries as they ripen and create enough weight to cause their red stems to droop.

You can find a palette of native elderberries growing in North America; American black elderberry (*S. canadensis*) is native to North America and grows in the East. Red elderberry (*S. racemosa*) is notably disagreeable eaten raw, and though Native Americans of the Pacific Northwest used it in food preparations, it's generally not preferred by today's elderberry lovers—most sources say they are toxic. Blue elderberry (*S. cerulea*) grows in western and central states, all the way up to Canada and all the way down to Mexico. Its berries are covered in a

powdery bloom. *S. nigra,* the European or common elderberry, is often what gardeners will plant for harvesting.

From Pacific to Atlantic, there are few regions without a variety of elderberry. They are not terribly picky about where they grow: partial shade, full sun, in forests. They often grow close to streams and creeks. I found a shrub growing in the drainage ditch between the parking lot and the road at my local grocery store, adjacent to some cattails. Since harvesting it in a spot that likely had a lot of questionable road runoff seemed a bad idea, I just appreciated it from afar.

Before the berries come the elderflowers: small, creamy white, and in broad clusters that can be wider than the span of your hand.

Lightly sweetened elderflower cordials and liqueurs taste of spring and new sunlight. A classic wild foods preparation is to dip the entire flower cluster in pancake batter and then griddle it to make elderflower pancakes. (Frying flowers coated in a tempuralike batter to make lacy fritters is another version.)

These pale, feathery delights could not be more different in personality than elderberries. As they mature, the fruits become dark and mysterious and brooding. Elderberries are not grazing berries. They taste flat and uncompelling to many when eaten raw. You can sample one or two just to see for yourself, but it's not advisable; raw elderberries are disagreeable to digestion, so don't eat too many. In fact, all parts of the plant, save the flowers and berries, are mildly toxic.

Elderberries are some of the littlest berries. Beady and nearly black when ripe, they remind me of the eyes of dead mice caught in a sprung trap. In contrast to producing such small berries, the plants themselves can be big—bona fide trees, in some cases. *S. nigra* can reach up to 20 feet (6 m).

The light but pliable stems are easily hollowed out. Elderberry branches were used to make spiles for harvesting the sap of sugar maple trees. Those handy branches account for the plant's scientific name: *Sambucus*. A sambuca or sambuke was a primitive woodwind made of an elder branch or stem. Fans of spirits may be curious about the Italian anise liqueur Sambuca and its place in all this. Sambuca contains elderflower, among other aromatics, and also owes its name to the elder. One theory speculates the common name derives from the Anglo-Saxon word *aeld*, which means "fire"; the hollow branches were useful for blowing air, like a mini bellows.

Elderberry juice can give cranberry juice a run for the money in its ability to stave off urinary tract infections. Its best-known medicinal use is to battle colds and flus, as either a treatment or a preventive.

Harvesting and Storage

Harvest the ripe berries in the late summer or early fall when they are dark purple (or blue and powdery, if it's blue elderberries in your neck of the woods). Wait until all of the berries on the clusters are ripe at once. Just cut off the entire cluster and pop it into a grocery sack to deal with once you get home.

Getting the berries off the stems is the part that may challenge your patience. There are a few approaches. One is to strip them from the heads using forks. Another is to freeze the clusters in bags overnight, then drop the bag onto the floor to jostle the frozen berries free. Before preparing the berries, submerge them in cool water and let any bits of stem float to the top so you can easily pick them out (remember, the stems are mildly toxic, so be thorough).

Elderberries start to break down soon after you pick them. Account for that before you even harvest them; you'll want to process them within 48 hours. Refrigerate or freeze the picked berries as soon as you can. If freezing them, be sure to rid them of debris first—once thawed, the juice from the berries will make spotting any stray bits of stem difficult.

Elderberries dry well, but if you use a dehydrator they may fall right through your racks. Put them on screens. You can try spreading them out and letting them dry passively, too. The dried berries are often used to make syrups.

If you are a new plant watcher eager to observe your neighborhood elder and snatch up its flowers in the spring, be especially careful with your identification. Elderflowers have a few look-alikes that you don't want to mess with, such as water hemlock (poisonous!) and red osier dogwood.

Culinary Possibilities

Elderberries are a raw material awaiting your creativity for transformation. Remember, you'll need to cook or ferment the berries for them to be safe to eat in any significant quantity. That points you to homemade wines and brews in one direction and baked goods, syrups, and preserves in another.

Before you get too far, it's good to know that of all the inky-dark berries, elderberry juice has a singular propensity to spatter on counters, clothing, and cupboards, leaving blotchy stains. Also, as they cook, elderberries on their own smell of scorched hot cocoa: not particularly alluring, I know, but not rank, either. I also think they taste and smell much better when combined with other ingredients.

Elderberries have an old-fashioned vibe to them, and I enjoy using them accordingly.

The strained juice of elderberries simmered in water is the foundation for syrups, cordials, and sauces. The juice itself is attractive, and elderberries have a high skin-to-flesh ratio, making their mouthfeel not particularly luscious when used whole.

Elderberry Kir Royale

Makes 1 cocktail

A kir is a white wine and crème de cassis aperitif. A kir royale gets an upgrade with Champagne and is therefore my preferred version. "I will drink Champagne anytime, anywhere, with anyone," was a famous saying of one of my wines instructors in cooking school. This is a recipe for fun or trouble, or (more likely) both. Since the forager's pantry likely includes mysterious bottles of home-infused fruit liquors and syrups, instead of buying crème de cassis, break out your Elderberry Cordial for this classy and celebratory drink.

Chilled Champagne or dry sparkling wine, as needed
½ ounce (15 ml) Elderberry Cordial (page 124)

Pour the Champagne in a flute glass. Add the Elderberry Cordial slowly. Party time!

ELDERBERRY CORDIAL

Makes about 1 quart (960 ml)

Cordials are heavily sweetened fruit syrups once sipped for their medicinal value. Cordials do not always contain alcohol, but some of them do. A certain type of bookish girl will recall cordial from a scene in L. M. Montgomery's *Anne of Green Gables* in which young Anne mistakenly gets her best friend Diana tipsy by serving her currant wine instead of raspberry cordial.

This cordial is free from tipple, and thus safe to serve to elderberry fans and best friends of any age. The procedure below is more of a template, so feel free to add a few cloves or use all lemons instead of lemons and oranges. I think elderberries need a little more help in the flavor department than other berries, but don't go overboard like you're making an Elderberry Spice Latte. This method of making the elderberry juice is from forager Hank Shaw, who says it results in a larger yield than simply boiling and straining the berries.

In olden times people took small glasses of elderberry cordial daily as a preventive for colds and flus. Doing so today certainly beats taking Emergen-C! If you are drinking it for pleasure and not as a tonic, try mixing it with fizzy water, sweetening hot tea with it, or drizzling it over ice cream.

5½ cups (910 g) elderberries
2 cups (400 g) granulated sugar
¼ cup (60 ml) honey
One 3-inch (7.5 cm) cinnamon stick
Zest and juice of ½ orange
Zest and juice of 1 lemon

Carefully sort through the elderberries, and remove as many stems as possible. Put them in a large bowl, and cover with water. Discard any little bits that rise to the top.

Drain the elderberries, and put them in a large nonreactive saucepan. Add just enough water to cover. Pulse them a few times with an immersion blender on low (you want to break up the skins somewhat, but not any seeds); alternatively, mash them a little with a potato masher. Bring to a boil, then reduce the heat to a simmer and cook 20 minutes.

Strain the mixture through a conical food mill and use the pestle to extract as much liquid as you can from the solids. Or pour the mixture though a jelly bag, letting it drain for about an hour and then squeezing with your hands to extract as much juice as possible.

The Limitations of Food as Medicine

Periodically, trendmakers and food marketers get all hopped up on the health benefits of specific foods. Since fruits often contain beneficial compounds and nutrients—antioxidants, fiber, anti-inflammatories—it seems every year a different one gets the royal treatment as a miracle superfood. "Our faith in the power of one ingredient—pomegranates, kale, Greek yogurt, acai—to save us, heal us, give us eternal life, perfectly mirrors our postwar cropping style: single-minded devotion to the One, and pure hatred for the Other," wrote Dana Goodyear for *The New Yorker*.

Just like people, foods are not exclusively good or evil. Foods have medicinal qualities, yes, but I don't like to think of food as medicine. It's part of the fabric of our lives, both a necessity and a pleasure.

We have the benefit of eating a variety of foods year-round, so a balanced and eclectic diet is part of a healthful lifestyle in our modern world. If I eat goji berries, it's because I feel like eating goji berries, and not because I think I'm supposed to. Fortunately, I often feel like eating healthful foods because I like the way they taste. Sometimes I eat fried bologna sandwiches because I like the way those taste, too. A fried bologna sandwich could reasonably be considered the opposite of a superfood, but it is Super Food.

The healing powers of food plants only work if there's a modicum of joy involved. Ideally, every meal itself is a prayer of thanksgiving for the marvel of having enough food to keep your body going another day or so. Remove that part and you might as well be swallowing pills.

Measure the juice: For every 4 cups, add 2 cups sugar, ¼ cup honey, 1 cinnamon stick, the zest and juice of half an orange, and the zest and juice of a lemon. Bring to a boil and cook for 5 minutes, skimming and discarding any foamy scum that rises to the surface. Taste the mixture once it's cool enough: it should be quite sweet and sticky. Cool, pour into sterilized glass bottles, and seal tightly. Refrigerate the cordial for up to 1 year.

LOOK
→ BUT ←
DON'T
EAT!

POKEWEED

Phytolacca americana
Phytolaccaceae family
Eastern US and Canada; western US

Parts of pokeweed are edible, but controversially so. Don't eat pokeweed casually. *Definitely* don't eat pokeweed berries, casually or not. A common sight in eastern North America, pokeweed is a fascinating plant with an abundance of lore, and that's what you should forage for: its history.

Pokeweed even has its own unofficial theme song, recorded by Tony Joe White in Muscle Shoals, Alabama, in 1968. That automatically makes it cooler than other weeds. The song is "Poke Salad Annie," and it's more or less about southern rural poverty. Young pokeweed leaves can be blanched multiple times and eaten as pot

greens—the poke salad (or poke salat) that the titular Annie fixed to feed her siblings, since that's all they could afford to eat. It was a food of lean times. Skip the leaf-blanching and you'll have a bellyache at best, a hospital trip or dead-end journey in a pine box at the very worst. The knowledge and effort required to safely forage and prepare this plant quite likely account for its disappearance from the dinner table, since it's a lot easier to buy a bag of pre-washed baby spinach from Whole Foods and quickly sauté that instead.

Pokeweed plants thrive in disturbed soil in sidewalk cracks, along riverbanks, and near cinder-block buildings and collapsing barns. When the weather gets hot, they grow almost exponentially, starting from green shoots in the spring to towering things over 6 feet (1.8 m) tall, with dark ruby stems and giant paddle-shaped leaves. Around August dark blue berries form on droopy tendrils from its top. The berries are attractive, and I like spotting them as I walk past the back alley compost piles and garbage cans pokeweed seems to favor as companions. The plant also grows merrily in the woods and near creeks and riverbanks.

Pokeweed seeds are poisonous, and the berries are not delectable. I've read that they can be made into pies, but this seems careless—why tempt fate when we can make pies with so many tastier, risk-free berries? If you want to mess around with pokeweed berries, consider making a dye or ink. In fact, pokeweed was called inkberry by soldiers during the Civil War, who used the juice to pen letters back home. (Note that another plant, *Ilex glabra*, an evergreen holly, also goes by the name inkberry.)

Pokeweed dies back in the fall, and its once richly colored stalks wither and brown in the cool weather, looking nothing like the plant that thrived so assertively in the summer. But it'll come back the next year, just as tenacious as ever, reminding you of the resourcefulness that carries people through trying times.

Pontack

Makes 1 half-pint jar (240 ml)

Who's up for making purple Worcestershire sauce? Heck yeah! Pontack is a quaint elderberry-based British condiment used on rich meats like game or pork, but you can add it to other recipes as you would Worcestershire. Folk wisdom advises not to use it until it's sat for 7 years. I love food projects that last longer than most relationships, especially when the vast majority of the investment is passive storage. Besides, you can crack open your pontack after as little as 3 months, if you must. If you were lucky enough to have a gangbusters elderberry harvest, scale this up.

2⅔ cups (440 g) elderberries

2 cups (480 ml) apple cider vinegar

2 large or 3 small shallots,
    peeled and sliced

One 3-inch (7.5 cm) piece fresh ginger,
    smashed a bit with the flat side
    of a knife

4 oil-packed anchovies, patted dry,
    optional

1 teaspoon whole cloves

½ teaspoon mace

½ teaspoon nutmeg

1 tablespoon black peppercorns

5 allspice berries

¼–½ teaspoon kosher salt

Preheat the oven to 250°F (120°C). Rinse and stem the berries. This will take a while. Put on one of your favorite records, and drink something refreshing and enjoy it. Place the berries and vinegar in a deep, ovenproof glass, ceramic, or enamel casserole. Stick it in the oven for 4 to 6 hours, or overnight. This will make your house reek, by the way.

Remove the mixture from the oven, and strain through a sieve—or better yet, a conical food mill—crushing the berries with a pestle or your hands to extract as much juice as possible.

Put the strained liquid in a medium nonreactive saucepan, along with the remaining ingredients. If you added anchovies, use ¼ teaspoon salt; if not, use ½ teaspoon. Bring to a simmer, and cook for 20 to 25 minutes, until slightly reduced. Remove from the heat, and strain through a cheesecloth-lined sieve. Discard the solids.

Return the liquid to the pan. Bring to a boil, and keep at a boil for 5 minutes. It should be slightly syrupy, but not particularly thick. Pour the sauce into a warm, sterilized glass bottle or jar, and seal. Label and date, and store in a cool, dark cupboard for at least 3 months before using. Let me know if you make it to 7 years.

NOTE: This stuff produces incredibly pungent fumes as it cooks. I made it one day when my daughter had a friend over to play, and they dramatically donned clothespins on their noses. If the weather is nice, make it outside: If you have a slow cooker, rig it up on a covered porch and use that to steep the berries in the vinegar. This will help maintain your popularity with any cohabitants, plus keep your house cooler.

FIGS

Ficus carica
Moraceae family
Throughout the US

IN 2004 THEN TWENTY-FIVE-YEAR-OLD Californian Alastair Bland decided to drop out of grad school and explore his home state via bicycle on a 2,000-mile (3,200 km) tour. During those two months, Bland traveled free of tent or any provisions to speak of: "I thought it would be exciting to try to live solely off of found and foraged food during my journey," he wrote. "I would only eat that which was available in public places or given to me freely, and I would get to taste all the natural flavors of California."

Though pears, grapes, almonds, walnuts, avocados, and prickly pears were some of those natural flavors, California's many renegade fig trees provided him with his primary fuel: "Figs are a rich, solid food, and even after a day of exertion, a dinner consisting of nothing else was filling."

Bland zeroed in on an excellent fruit to sustain his odyssey. Figs are nutrient-packed, offering more minerals per serving than other common fruits. They're also high in fiber and sugar, keeping people on the move in multiple senses.

Where there is mythology and lore—whether ancient or contemporary—there is the fig. Banyan trees, those droopy-rooted trees with trunks that seem to melt over jungle temple ruins, are *Ficus*. In the Quran the Prophet Muhammad said, "Eat figs! If I would say a certain type of fruit was sent down to us from the heavens I would say it's a fig." At least forty-four verses in the Bible mention figs.

Only a few species in the genus *Ficus* produce fruit palatable to humans. Around one-tenth of all bird species and one-sixth of all mammals eat figs. The trees are native to western Asia and were one of the first plants to be domesticated. It helps that high temperatures and drought are no big deal to a fig tree. There even are some hardy varieties of figs that can withstand cold winters, and those are precisely the kind growing two blocks away from my house (the county where we live is currently classified as USDA hardiness zone 6a). Whoever is growing those little fig trees loves figs, because they pick off every one right after it ripens. Trust me, I keep tabs. Those figs are the figs of my own mythology.

The anatomy of a fig is contrary to that of most other fruits, because a fig is not itself a fruit. In a very feminine gesture, the blossoms do not appear on the tree, but inside the fig. It is a pod surrounding the *achenes*, or seeds, each of which is its own individual fruit. Consuming a single fig is eating hundreds of fruits at once. No wonder they are so filling.

With all of these flower/fruit/seeds tucked discreetly inside the fig, how does it get pollinated? The answer is—for wild figs at

least—wasps. Figs and wasps co-evolved, and in this symbiotic relationship tiny, ant-sized female fig wasps lay their eggs in figs. They enter through the *ostiole*, the opening at the bottom of the fig, and their wings and antennae often break off as they do so. The larvae hatch and mate inside the fig, then either die inside or bore a hole and exit to continue the cycle. After this, the fig ripens and becomes alluring to other animals. If this piece of trivia now makes figs unalluring to you, don't fret. Many figs these days are self-pollinating and make it to you with no evidence of wasp reproduction.

Figs arrived in the New World via the Spanish, who were responsible for unleashing all kinds of delightful fruits on the Western Hemisphere. The Spanish Jesuit missionary and naturalist José de Acosta, who visited Mexico in 1586, listed figs, pomegranates, olives, and mulberries among the introduced plants growing there. Missions loom large in the tale of figs in North America, as does California. The land supplies figs with the Mediterranean climate they adore, and black mission figs are thus named because the gardens at Mission San Gabriel (in what is today Los Angeles County) included fig trees. Mission figs were *the* figs in America until immigrants from Europe and Asia brought cuttings of other varieties with them to propagate. The

yards of what are or were immigrant neighborhoods are potential gold mines of lesser-known fig varieties.

Shifting to the East Coast, Ocracoke Island in North Carolina's Outer Banks is a hotbed of fig culture. Isolated geographically and doggedly small scale, it's a beloved summer vacation spot with a year-round population of under a thousand residents. The number of fig trees on the island was not documented in the most recent census, but it's a good thought. Figs trees have likely grown on the island nearly as long as it has been settled, since the eighteenth century. Some of the fig trees on the island are over a hundred years old. Ocracoke has an annual fig festival, a highlight of which is a fig cake contest. Ocracoke fig cake uses preserved figs, an island specialty, as a base. It's like moist and spicy carrot cake, yet better, because it has figs in it.

Figs hold symbolic importance to millions of people who don't even eat them or could not identify a tree (via the Bible, et cetera); the concept of figs unifies human beings. Living figs, meanwhile, are a fruit for dynamic people who like having all of their sensory receptors hit. Either way, whether you eat them or not, figs are part of you. They are our past and future all in one bite.

Harvesting and Storage

Fresh figs spoil fast. Refrigerate them (in a single layer, if possible) to coax a little longevity out. You have a window of 3 to 10 days, max, before they'll get soft and spotty with mold.

Because they are not built for long-term shipping or storage, those of us who do not live near productive fig trees are most familiar with dried figs . . . or, dare I say, Fig Newtons. But if the gods have blessed you with proximity to figs, you can capture the fleeting Season of the Fig by canning, baking, and dehydrating.

You can freeze figs, too. Like delicate berries, try freezing them on a tray until firm before packing into bags or containers. Use the thawed figs in applications where they'll be cooked, as the freezing and thawing will damage their structure.

Culinary Possibilities

A good, ripe fig offers a panoply of textures: crunchy seeds, tender skin, sweet and creamy flesh strung with a network of fibers. When you find exemplary specimens, eat them up on the spot.

Figs excel in savory applications, perhaps because they have a meaty texture. Quarter them and use them as a topping for white pizzas, roast them with rich meats, or sauté them with sturdy cooking greens. Halve them, fill their centers with blue cheese, and drizzle them with honey or a good balsamic. Or wrap quartered figs in slices of prosciutto, perhaps flashing them on the grill first.

After doing all that, it is possible you still have buckets of fresh figs. I vote to make preserved figs and can them, which will be your ticket to year-round figgyness.

Interestingly enough, don't go dropping raw figs in your Jell-O mold. Figs contain proteases, which are enzymes, and they interfere with the collagen in the gelatin setting. Jell-O desserts get a bad rap, but I am sure you can easily think of better things to do with your figs, anyway.

Ocracoke Island Fig Cake with Brown Butter Glaze

Serves 10–12

If Ocracoke Island had a cake mascot, this would be it. Many recipes for this moist spice cake exist, but they are all essentially the same recipe with assorted small adjustments. The ingredients are more or less fixed. The amount of fig preserves seems to dictate the ratio of sugar and buttermilk. Also, all the recipes I've read call for vanilla extract, which I omit. I've never been to Ocracoke, but I've been obsessed with going there in August when the figs are ripe ever since I learned of the island's fig culture. Maybe if I ever make it there I'll start adding the vanilla.

Glazes and frostings are not fixed. Some cakes are au naturel, but since a drizzle of glaze elevates a cake from appealing to irresistible, I recommend throwing this one together.

For the cake
3 eggs
1 cup (200 g) granulated sugar
1 cup (240 ml) vegetable oil
2 cups (260 g) unbleached all-purpose
    flour, plus more for dusting the pan
1 teaspoon baking soda
1 teaspoon cinnamon
½ teaspoon freshly grated nutmeg
½ teaspoon ground cloves
1 teaspoon table salt
½ cup (120 ml) buttermilk
1½ cups (360 ml) roughly chopped
    Preserved Figs (page 138, don't
    include the citrus slices, if there
    are any)
1 cup (120 g) chopped pecans, toasted

Preheat the oven to 350°F (175°C). Grease and flour a 10-inch (25 cm) Bundt pan or tube pan. Set aside.

In a large bowl, beat together the eggs, sugar, and vegetable oil until thick.

In a medium bowl, whisk together the flour, baking soda, cinnamon, nutmeg, cloves, and salt.

With a rubber spatula, fold half of the flour mixture into the egg mixture. Then fold in the buttermilk, followed by the remaining flour mixture. Fold in the figs and nuts just until combined.

Scrape the batter into the prepared pan. Bake until the top springs back when you press it with your fingertip and a wooden skewer inserted in the center of the cake comes out clean, 40 to 60 minutes. Set on a wire rack to cool for 15 minutes.

For the glaze

¼ cup (½ stick) unsalted butter,
  cut into 4 pieces
1 cup (115 g) confectioner's sugar
1 tablespoon freshly squeezed
  lemon juice
1 tablespoon heavy cream
Pinch salt

As the cake cools, make the glaze: In a medium skillet over medium heat, melt the butter. Once the foam subsides, whirl the skillet around every minute or so to agitate the butter. Keep a close eye on it; eventually the creamy white bits (the milk solids) at the bottom of the pan will turn a nutty brown, and the butter will have a nutty aroma, too. This could take 5 to 6 minutes. Right when the butter is brown, whisk in the powdered sugar, lemon juice, cream, and salt until smooth. The glaze should be drizzle-able, but not too thin. Set aside, and keep warm.

Loosen the sides of the cake from the pan with a thin metal spatula, if necessary, and unmold the cake. Set it on a serving plate or platter. If the glaze has separated, whisk it briefly, then pour it over the warm cake. Let the cake cool completely before serving. The cake will keep, covered, for 3 days.

Pork Tenderloin with Rosemary Roasted Figs and Onions

Serves 4–6

Fresh figs are a summer food, and this dish is pretty heavy eating for a hot summer night, but sometimes it's best to go big. If you can bear cranking up the oven, this is worth it—it's nearly a one-pan dinner. Make some polenta or roast up some potatoes to go on the side, and you're ready to go.

1 (1½–2 pound/680–910 g)
    pork tenderloin
Salt and freshly ground black pepper
4 sprigs fresh rosemary
12 fresh figs, any variety
1 large red onion, peeled and
    cut into wedges
2 tablespoons extra-virgin olive oil
2 teaspoons balsamic vinegar
½ cup (120 ml) dry red wine
2 tablespoons unsalted butter

Preheat the oven to 400°F (200°C), and position a rack in the center.

Pat the tenderloin dry with paper towels. Season generously with salt and pepper. Lay the rosemary sprigs in the bottom of a metal roasting pan or large (12-inch/30 cm) ovenproof skillet, and set the tenderloin on top. Arrange the figs and onion wedges around the pork. Drizzle the whole works with the olive oil and balsamic vinegar, then sprinkle the figs and onions with salt and pepper. Roast until an instant-read thermometer inserted in the center of the tenderloin reads 155°F (68°C), 30 to 40 minutes.

Remove the pork, figs, and onions to a serving platter; tent with foil to keep warm, and allow the pork to rest for 10 minutes.

Meanwhile, make the sauce: Discard the rosemary. There should be a decent amount of liquid in the pan. Set it on a burner over high heat, and deglaze with the wine. Boil until reduced to about ¼ cup (60 ml). Turn the heat down to medium-low, and add the butter, swirling the pan until the butter is completely melted and emulsified into the reduction. Taste the sauce, and adjust seasonings with salt and pepper, if needed (there won't be a ton of sauce, but it should be very highly flavored). Keep warm over low heat.

Slice the tenderloin thinly on the bias and against the grain, then arrange on a platter or serving plates. Drizzle the sauce over the top, and serve immediately.

VEGETARIAN OPTION: You can play with the same flavors and just not use meat. Stewed *lentilles de puy* are a great accompaniment to this. Omit the pork, use a smaller roasting pan, and follow the procedure above, adding ¼ cup (60 ml) vegetable stock along with the wine if the liquid in the pan after roasting seems scant. Garnish the figs and onions with ½ cup (60 g) toasted walnuts and, if you like, ¼ cup (28 g) crumbled blue cheese.

GRILLED OPTION: You won't get the nice sauce, but this is fun for grilling fans and very elemental in an out-of-doors cooking fashion (if you can find some, thick-cut pork chops instead of the tenderloin are great). Strip four sturdy rosemary sprigs mostly of their leaves and thread the figs on them (you may alternatively use soaked wooden skewers). Keep the red onion in halves or quarters. Rub the pork, onions, and figs with olive oil, and sprinkle with salt and pepper. Grill over high heat until the center of the tenderloin reads 155°F and the figs and onions are browned in spots (the pork, figs, and onions may finish at different times). Just before serving, drizzle a little balsamic vinegar over everything.

Roasted Kale with Figs
Serves 4

As simple as it gets. I'm not a fan of raw kale, even when it's been massaged lovingly with flavorful dressings. The closest I get is this mess, which is a wilted kale salad / kale chips hybrid. It's inspired by a failed batch of kale chips a friend served at a birthday party—she'd dumped the kale leaves in a big pile on the sheet pan, so half the kale steamed instead of crisping up in the oven. The result was oily and salty, and I think I ate half the bowl. Adding fresh figs offsets the bitterness, and they get steamy and a little slouchy as they roast. If your oven has a convection setting, use it, and you'll get really appealing crispy edges on some of your kale leaves.

1 bunch kale, stemmed and torn
    into large pieces
4–6 fresh figs, any variety, quartered
3–4 cloves garlic, sliced crosswise
    very thinly
2 tablespoons extra-virgin olive oil
1 teaspoon sherry or apple
    cider vinegar
Salt and pepper to taste

Preheat the oven to 400°F (200°C). Position a rack in the center.

Tumble all the ingredients together on a rimmed sheet pan until the leaves are somewhat evenly slicked up with the oil. Season generously with the salt and pepper. Roast 5 minutes, toss, and roast until the outer edges of some of the kale leaves start to brown, 5 to 7 minutes longer. Serve hot or at room temperature.

ALSO TRY WITH: A firm, thinly sliced apple is nice with this.

Preserved Figs

Makes 2–3 pints (960 ml–1.4 L)

Canning figs is a tradition of Ocracoke Island in North Carolina's Outer Banks, where fig trees grow in many of the small island's yards. Depending on the person who cans them, preserved figs can range from jammy and goopy to intact and gently stewed in syrup. These are the latter style. Use any variety of fresh fig you like; brown turkey figs and celeste figs are common on Ocracoke.

These preserves are the beginnings of dense, fruity Ocracoke Fig Cake, but they can do so much more. Try spiking them with spices to make *mostarda* to serve with roasted meats, or spooning them onto yogurt or pound cake.

8 cups (910 g) figs
1 cup (200 g) granulated sugar
1 lemon, thinly sliced
1 orange, thinly sliced
1 cup (240 ml) water

If you like, stem the figs. They add some nice texture; most Ocracokers don't stem the figs.

In a Dutch oven or large nonreactive saucepan, toss the figs with the sugar and lemon and orange slices. Cover, and refrigerate overnight.

The following day, add the water. Cover, bring the figs to a simmer over medium heat, and reduce the heat to a low simmer. Cook, covered, for 1 hour.

Remove the lid, and continue simmering very gently for another 1 to 2 hours. You want the syrup to thicken, some of the figs to break down a bit, and the citrus rinds to get somewhat translucent.

Meanwhile, sterilize your jars, and prepare a bath for water bath canning. When the figs are stewed to your liking, transfer them, along with the citrus slices and syrup, to the hot jars, leaving ¼ inch (6 mm) headspace. Screw on the lids so they are fingertip tight and process in the water bath for 5 minutes.

NOTE: You can scale this up, but don't go crazy; if you have a ton of figs, it's better to can a few batches. If you'd like a more syrupy version, double the sugar.

GOOSEBERRIES

Ribes spp.
Grossulariaceae family
Throughout the US and Canada

GOOSEBERRIES ARE HAIRY AND THORNY, while their cousins, currants, are smooth and free of prickers. But there are many species of gooseberries growing in North America, some more worthy of risking a prick than others. Pies, fools, syrups, and wines await those who opt to seek out what all too many fail to notice.

Failing to notice gooseberries is easy to do. What presence in our daily lives do gooseberries have? Even the domesticated varieties are uncommon in our North American fruit vernacular. This is in part because of government planting restrictions put into place in the 1930s on currants and gooseberries to curb the spread of white blister pine rust (many of those bans are now lifted on gooseberries), but I think it's cultural, too. Calling gooseberries our own is not a thing we do here.

We have plenty of native gooseberries to call our own when we get around to it. Take Canadian gooseberry, Idaho gooseberry, Sierra gooseberry, desert gooseberry, and Appalachian gooseberry, to begin with. Their names will give you a general indication of their ranges. Gooseberries thrive in cool areas with moist soil, but some varieties can be found in arid areas. Besides having a variety of habitats, they have a variety of appearances, from tiny and spiny to smooth and veiny. But they all grow on shrubs with lobed leaves bearing a somewhat maple-ish shape. The berries can have a pronounced papery brown

tail hanging off their bottoms. That's what's left of the flower from whence they grew. Before those flowers shrivel up, you will notice them dangling like a garland in a line from the branch. Currants, meanwhile, dangle in clusters.

Some unripe gooseberries seem to be tiny props from a cheesy science-fiction movie: They're spiky, like a menacing alien life-form on a strange planet. Harvest these prickly guys with gloves. If a gooseberry is fully ripe yet looks too spiky to pop straight into your mouth, it is. You'll need to cook it, allowing you to access the tasty pulp. Smooth gooseberries may be enjoyed straight from the bush, if they are sweet enough.

Even ripe gooseberries can be on the tart side. American gooseberries are usually smaller than European varieties, which can be up to 1 inch (2.5 cm) long. The plumper and spineless European varieties are generally recognized as easier to cultivate and better tasting, though I've not sampled the two side by side. I say if you find gooseberries at all you should be content with the ones you find.

While some discover their love of gooseberries in the Sierras, legendary cookbook editor Judith Jones wrote about tasting her first ones in Wales, where they were served in a sauce over mackerel: "The tartness of the gooseberries cut through the oiliness of the fish in a very alluring way." Smitten, she and her husband planted bushes at their Vermont summer home after returning from their trip abroad. European gooseberries have naturalized in some areas of the Northeast, though I don't think this can be traced back to Judith Jones's vacation house.

The Brits are hands down the biggest gooseberry-philes in the world. The classic recipes of the gooseberry canon reflect this: gooseberry tart, gooseberry fool, gooseberry crumble, gooseberry flummery (a chilled, cooked porridge of fruit). They also have a few dusty old gooseberry sayings (as far as I can tell, we have zero). To "play gooseberry" is to chaperone a pair of young lovers. To find a baby "under the gooseberry bush" is a euphemistic way of explaining the arrival of a new infant to young children—a fruity version of the stork. "Old Gooseberry" is slang for Satan himself, though it's not a very intimidating nickname, is it?

For a period in England there was a mania for competitive gooseberry growing. Gooseberry societies had contests for the largest gooseberry, and in 1845 there were 171 such shows. Though the hobby has fallen out of fashion, it still has pockets of enthusiasts. The world record set in 2013 by Kelvin Archer of Cheshire was with a berry tipping the scales at 64.49 grams. That may sound small, but compared with a wild gooseberry it's like Jupiter is to Earth. The gooseberries you encounter will be puny in comparison, probably like slightly bigger currants.

Kiwis were once marketed in America as Chinese gooseberries, but they are not related (they are, however, native to China and not New Zealand). So-called Cape gooseberries are likewise not gooseberries, but ground cherries.

Harvesting and Storage

Not all gooseberries have thorns, but wild ones likely will. Try to be deft when you pick them. They should come off the plant easily when they are ripe.

When you pick gooseberries may depend on what you want to do with them. To eat out of hand (if that's an option, given the possibility of thorns or hairs), you want to seek the ripe ones. For cooking, slightly underripe gooseberries are preferred. Figure this out simply by tasting the berries from time to time and picking ones with the attributes you like.

Color is not quite an indicator of ripeness, as gooseberries vary in color from white to green to red to blackish. If you'd like to be precise about picking, harvest them one by one; if ripeness is no matter or you don't mind sorting through your harvest to discard leaves, stems, and underripe berries, wear thick gloves and use your hands to strip the branch of its berries into a container.

Culinary Possibilities

For jelly, try underripe berries because they contain sufficient pectin to gel on their own, whereas ripe berries require adding commercial pectin. Mix ripe and underripe berries together in jam, as Jones does, for the best of both worlds: a preserve with nice blush that sets well and has a rounded flavor.

Hairy gooseberries need to be cooked and strained to be enjoyed. The resulting juice can be a base for syrups, wines, and creatively constructed beverages. Gooseberry shrub? Why not?

Take advantage of the tart and lemony taste of gooseberries in recipes where you'd use lemons, particularly with oily and assertive fish or meats like veal or pork. Cook the berries down with a pinch of sugar and you have a nearly instant chutney with the appropriate chunky-saucy body.

Infinite Bucket List

"Most people know the names of fewer than five wild plants that grow in their area," I read in a newspaper clipping. I have no idea where this statistic initially came from, or who these people are, but I can believe it. I'd bet most people *think* they can't name five wild plants, but if pressed could at least muster two or three. The view from any given office or classroom window probably affords one a glimpse of five wild plants, even if—or especially if—the window overlooks a dismal parking lot. That's not to say any five plants in a parking lot are widely thought of as exciting, because they'd likely be weeds, and people don't love weeds, or even think of them as plants.

This lack of plant literacy might seem dismal, but it's actually promising, rife with possibilities. If I went outside right now, I could name five wild plants I saw, but there'd be a dozen more I'd not know. You can spend a lifetime learning about plants and still have new things to discover every day. For a curious but very part-time forager, the opportunities to sate your thirsty brain with exciting observations are infinite. What are those plants? Where do they like to grow? What animals do they attract? What seasonal cycles do they follow? Can you eat them? If so, do they taste good enough to bother with? What can those plants do for you, and vice versa?

The concept of creating a list of sights to see or thrilling activities to do before you croak is one that deeply resonates with consumers—you can tell from the titles of hundreds of travel books and articles. *133 Experiences of a Lifetime. 1,000 Foods to Eat Before You Die. 51 Places to See in India Before You Die.* What pressure! Let's say you visit, eat, or do all of these things. Then what?

I have dreams and goals, too, but to begin with, one of them is to explore where I live, a place thought of by the general public as largely unexceptional (it could be a book: *1,000 Clichéd Ways to Dis the State of Ohio*). What an awful thing, to waste day upon day yearning for faraway places while the small wonders right under our noses go unnoticed. Acquiring even a middling passion for plants unlocks a world that's more easily accessed than you'd guess, but impossible to exhaust. Five wild plants is just a beginning.

TROUT WITH GOOSEBERRY SAUCE

Serves 4

In a fantasy world, you'd make this with trout you'd caught in a pristine mountain stream and gooseberries you'd plucked from bushes growing at a rustic country cabin that's been in the family for generations.

My trout came from the fish counter, but this combination of rich fish and tart, citrusy gooseberries is still dreamy. You can also serve the sauce over other oily fish, like mackerel, or over roasted or grilled pork loin.

1 cup (150 g) green (underripe), nonhairy gooseberries, topped and tailed
2 tablespoons water
2–2½ tablespoons granulated sugar
3–4 whole trout, skin on, filleted
Salt and freshly ground black pepper to taste
1 cup (130 g) all-purpose unbleached flour
3 tablespoons unsalted butter

Put the gooseberries, water, and 2 tablespoons of the sugar in a medium saucepan. Bring to a simmer, and cook gently, stirring from time to time, until the berries have mostly broken down, 5 to 15 minutes. Taste, and add up to ½ tablespoon more sugar, if needed, but keep in mind the sauce will taste sweeter once it's on the fish. Set the sauce aside.

Pat the trout dry with paper towels. Season both sides with salt and pepper.

Put the flour on a deep plate or pie tin. Dredge the fillets in the flour, and tap off the excess.

Melt the butter in a large skillet over medium to medium-high heat. Once the butter stops foaming, add half the trout fillets skin side down, and cook without disturbing until the skin sides are golden brown, 5 to 7 minutes. Carefully flip the fish over, and cook the other side until browned, 2 to 3 minutes. Remove the cooked fillets to a plate, and tent loosely with foil. Add the remaining fillets to the pan, and cook in the same manner.

Spoon a little of the warm or room-temperature sauce over the hot fish, and enjoy, with more sauce on the side.

ALSO TRY WITH: Add 1 tablespoon very thinly sliced fresh sage leaves if you'd like to serve the sauce over pork.

Gooseberry Fool

Serves 6

This isn't going to win any beauty awards, but what gooseberry fool lacks in color it makes up for in appeal. It's way easier to make than a lemon mousse but delivers all of the pucker-sweet pleasure. Don't make this recipe with hairy/thorny gooseberries.

3 cups (450 g) green (underripe) nonhairy gooseberries, topped and tailed
¼ cup (60 ml) water
¾ cup (150 g) granulated sugar
1 cup (240 ml) heavy cream

The day or morning before you want to serve the dessert, combine the gooseberries with the water in a medium saucepan. Bring to a boil over medium-high heat, reduce to a simmer, and cook until the berries are saucy and completely broken down, 8 to 15 minutes (thawed frozen gooseberries will take the lesser amount of time).

Add the sugar, return to a boil, and cook until the sugar dissolves, 1 to 2 minutes. Scrape into a heatproof bowl, cool to room temperature, and refrigerate until cold, at least 2 hours.

Reserve ¼ cup (60 ml) of the cold gooseberry mixture for garnishing.

In a large chilled metal or glass bowl, whip the cream to soft peaks and fold in the remaining cooked gooseberries. Divide among six serving glasses, then top off the center of each with a dab of the reserved gooseberry mixture. Serve immediately, or cover and refrigerate for up to 4 hours. (Be mindful that the fools will absorb any strong aromas, like raw fish, in your fridge.)

GRAPES

Vitis spp.
Vitaceae family
Throughout the US; eastern and central Canada

GRAPES HAVE EXERTED MORE INFLUENCE on Western culture than any other fruit. First and foremost is wine, obviously, and its many symbolic and ritualistic manifestations in religion, food, and art. Pointing out the many pleasures of raisins, grape jelly, and Waldorf salad seems feeble in comparison.

But let us consider our grapes of North America in their own right, undimmed by our long and persistent romance with Old World grapes and their expertly fermented juices. Grapes grew wild coast to coast before European colonists arrived, and grapes grow wild still. Some are tiny and relatively useless as fruit; others are big ol' honking things, each as bulbous as the nose of a Muppet. They are dizzying in diversity and in history.

Most of the grapes we pay any mind are the table grapes found in zip-top bags at any grocery store, any time of year: green, red, sometimes black. I grew up playing in the woods, and it was only a few years ago that I realized actual grapes could—and sometimes do—grow on those feral vines snaking ubiquitously up local trees and across shrubs. A lot of them are piddly little pathetic grapes, but grapes they are.

Along backyard chain-link fences or on ersatz clotheslines, a small contingent of grape fanatics grow their own grapes every year, varieties that hark back to native grapes most Americans aren't aware of at

all. I have long dreamed of moving into such a house, with its own ready-made grape arbor, clustered with bunches of sweet and juicy grapes for snacking, canning, and relishing. Here are some of the native grapes that have endured the Cab-ernization and Thompson-Seedless-ization of our grape lexicon.

CONCORD

V. vulpina

Fox grapes (*V. labrusca*) are native to eastern North America. They are the parent of cultivars such as Catawba, Niagara, and others that were major players in American wine making before Prohibition. Many still carry on, exhibiting the hint of musky aroma they're known for, including the Concord.

A stone marker (suspiciously shaped like a tombstone) at Grape-vine Cottage in Concord, Massachusetts, reads: "Ephraim Wales Bull planted the seeds of a wild Labrusca grape found growing on this hillside which after three generations through his work and wisdom became in this garden in September 1849 the Concord grape." Bull

was trained as a goldbeater (one who makes gold leaf), but his passion was horticulture. He planted grapevines and began experimenting with varieties that could withstand New England's harsh climate. It took thousands of seedlings and graftings, but one specimen had the promise of functioning as both a wine and table grape and grew easily where other grapes would not. It was the Concord.

Bull did not profit from his development, but the grape made a giant impression. "Introduced in 1853, it so dominated American grape growing thereafter that it still defines the idea of 'grape' for most Americans," writes Thomas Pinney in *A History of Wine in America*. While too low in sugar to be an ideal wine grape, its color, range, adaptability, and productivity made it *the* grape to grow. Outside of California, it is still the most widely cultivated grape in America.

Finding Concord grapes fresh is another matter. The willingness of their flesh to separate from their skins ("slip skin," a hallmark trait of *V. labrusca*), their seeds, and their tougher skins are not what today's palates are looking for. We get our Concords in liquid or jellied form. Fresh Concords are a joy to cook with if you can get them, though. If you can't, whether it's a shot of Welch's at Communion or a glass of Manischewitz at a Seder, consider how Bull's developments shaped the concept of what grapes are for an entire nation.

MUSCADINES AND SCUPPERNONGS

V. rotundifolia

Muscadines greeted European explorers and were the first grapes cultivated in North America. Captains Philip Amadas and Arthur Barlowe, part of Sir Walter Raleigh's first New World expedition, reported in 1584 that the North Carolina coast was "so full of grapes . . . on the sand and on the green soil, on the hills as on the plains, as well as on every little shrub as also climbing towards the tops of tall cedars . . ."

Spanish missionaries in Florida were making muscadine wine as early as 1565.

Their name derives from their aroma, which calls muscat grapes to mind. Most wild muscadines bear dark fruit, but southerners refer to pale (bronze) muscadines generically as scuppernongs, after the Scuppernong River in North Carolina. The Mother Vine, a scuppernong vine growing on Virginia's Roanoke Island, is credited with being the oldest cultivated grapevine still growing in the United States.

Hearty and good for eating and wine making, muscadines are likely the grapes you will see growing in southern backyard arbors. "Native muscadine, or bullace, grew profusely all over our farm, particularly along the fencerows. The scuppernong was its domestic equivalent, although scuppernongs were light colored and tended to be sweeter. Like most families, we put up jelly from the muscadines and ate the scuppernongs fresh," Jimmye Hillman writes in "What We Ate Back Then," an essay about growing up on a "semi-subsistence" farm in south Mississippi during the Great Depression.

Scuppernongs are bulbous and intoxicatingly fragrant. Pop one in your mouth, suck out the gelatinous pulp, spit out the slippery seeds, and throw the thick and leathery skin to the side. There's good flavor in the skin, though it's followed by a prickly grip. Some folks just eat the whole deal.

LESSER WILD GRAPES

I fondly recall a grapevine that entwined a little stand of trees at the foot of a popular trail in Mount Tabor Park in Portland, Oregon. I ran there often, and when the grapes appeared—they were marble-sized and green when ripe—I usually plucked a few for snacking. Between other keen-eyed human visitors and whatever fauna frequented the park, the vine was stripped of grapes in a week or two.

All wild grapes are edible, but not all of them are tasty. The ubiquitous grapevines that grow wild in the woods and on fences where I live in Ohio produce tiny, rock-hard grapes that appear to completely skip ripening in favor of shriveling up to sad, dry brown pellets. They barely even look like grapes, at least the ones we're used to seeing. One hot and wet year, some finally ripened to a lovely purple shade. They still tasted like shit.

Harvesting and Storage

Grapes only ripen on the vine. Vintners use refractometers to help determine when to harvest, but you'll have to stick with your eyes and taste buds. Try grapes from different clusters on the vine to get a good sampling. When the grapes taste sweet and are nice and plump, cut off the entire cluster with pruning shears. Refrigerate the unrinsed grapes, still on their stems, for as long as a month, removing any grapes that get squishy and rotten from the bunch to keep the other ones fresh.

Culinary Possibilities

We will here address unfermented uses for grapes. Slimy seeds and leathery skins make many native grapes unrealistic for taking along to work as a snack, unless you relish grossing out your co-workers with your table manners. Likewise, dehydrating is problematic, because your results may resemble no raisins from your memory. Baking with native grapes can involve seeding and peeling, a time suck for only the very committed.

Grapes and preserves are soul mates. A friend of mine makes juice from their Concord grapes and cans it, adding a few whole grapes in each jar to use as garnishes for refreshing drinks. Grape jelly, meanwhile, is the classic American jelly because (a) it tastes amazing, and (b) it allows the cook to forgo seeding or peeling.

Sturdier grapes (seeded, if necessary) are excellent additions to rich braises, particularly poultry. Throw in some forceful herbs like rosemary or thyme. Quickly sauté tart grapes to garnish broiled fish fillets. Or just pile a cluster next to a grouping of good cheese and bring it to a party.

New World Grape Jam

Makes 2–3 half-pint jars, or 5–6 quarter-pint jars (480–720 ml)

Grape jam has a little more oomph than grape jelly. It'll make you reconsider its applications beyond the traditional PB&J. Make this with the bulbous, sweet, thick-skinned grapes we can proudly call our own as North Americans: Concord, muscadine, scuppernong. Squeezing the pulp from the skins increases your yield, but more important, it increases the grapey flavor. The technique of pureeing the skins with sugar before cooking the jam came from a recipe in the late, great *Gourmet* magazine.

3 pounds (1.4 kg) Concord, muscadine, or scuppernong grapes
3 cups (600 g) granulated sugar, divided
1 tablespoon lemon juice

Halve the grapes, if necessary. Pop their insides (the pulp) into one large bowl, and drop their skins into another large bowl.

Put the skins in a food processor or high-powered blender, along with about ⅔ cup (130 g) of the sugar. Process until liquidy. Scrape into a heavy-bottomed Dutch oven or large heavy-bottomed, nonreactive saucepan. Add the lemon juice, the remaining sugar, and the grape pulp. Stir, bring to a boil over medium-high heat, reduce to a simmer, and cook 20 minutes, stirring occasionally and skimming off the foamy scum that forms along the edges of the pot.

Transfer the whole hot mess to a food mill and force it through. Discard the scraggly solids left in the mill. Return the liquid to the pot, bring back to a simmer, and cook until the jam passes the gelling test or reaches 220°F (105°C) on a candy thermometer. This could take 20 to 45 minutes.

Ladle the jam into the hot jars, leaving ¼ inch (6 mm) headspace. Wipe off the rims, set clean and dry lids in place, and screw on the bands so they are fingertip tight. Process in a water bath for 10 minutes. The jam is best after its flavor has had a few days to bloom. Dole the jars out carefully, because they are gorgeous and their contents are amazing.

NOTE: If you don't have a food processor or high-powered blender, chop the grape skins finely with a knife and add all of the sugar to the pot when you start cooking the jam.

CHICKEN BRAISED WITH WILD GRAPES, MUSTARD, AND GARLIC

Serves 4–6

Scuppernong and muscadine grapes' leathery skins aren't easy to eat raw, but when you cook them they become surprisingly supple and melty. They add unexpected body to the sauce in this braise, which is off-sweet and tangy from the mustard. There's something very old-school about the combination of grapes and chicken, something very 1960s dinner party, but trust me: It's magnificent. You can make it a day in advance and warm it up to serve for a swanky dinner with wild rice or couscous to soak up the fabulous sauce. A side of quickly sautéed shredded bitter greens is a great foil for the grapes.

1 whole chicken (4½–5 pounds/ 2–2.25 kg), cut into 8 pieces
Kosher salt and freshly ground black pepper
¼ cup (60 ml) olive oil
1 large head garlic, separated into cloves and peeled
1 cup (240 ml) dry white or rosé wine
1 tablespoon whole-grain mustard
1 bay leaf
About 2 cups (300 g) scuppernong or muscadine grapes, halved and seeded
1 tablespoon fresh thyme leaves

Up to 4 hours in advance, pat the chicken pieces dry with paper towels, season them generously with salt and pepper, set them on a baking pan, and refrigerate them, uncovered. This helps the chicken brown better, but you can season the chicken right before cooking, if that's how your day is going.

Heat a large (about 5 quart/5 L) Dutch oven over medium-high heat. Add 2 tablespoons of the oil. Once it shimmers, add the chicken skin side down, being careful not to crowd the pan (do this in two batches, if necessary). Cook until the skin is golden brown, then flip the chicken and brown the other side. Remove the browned chicken to a large plate. Add the remaining oil, if needed, and brown the second half of the chicken.

Add the garlic, wine, mustard, and bay leaf to the Dutch oven. Bring to a boil, and use a wooden spoon to scrape the tasty browned bits from the bottom of the pan. Add the browned chicken, return to a boil, reduce the heat to a simmer, cover, and cook 10 minutes. The liquid will not cover the chicken, but that's okay.

Add the grapes and thyme to the pan, rearranging the chicken pieces with tongs so the grapes are submerged in the liquid and the chicken pieces sit on them. Simmer, uncovered, for 30 minutes. Some of the grapes will slouch out of their skins—don't worry about it.

NOTE: I break my chickens down for braising this way: two drumsticks, two thighs, two wings, and two breast halves, each of which I cut into two pieces because chicken breasts are often gigantic. This makes a total of eight pieces. If you don't want to break down a whole chicken, you may use about 3½ pounds (1.6 kg) bone-in and skin-on chicken thighs and drumsticks.

For the sake of time, I brown the chicken all at once: half in the Dutch oven and the other half in a heavy-bottomed skillet. I deglaze both pans with the wine, then pour the liquid from the skillet into the Dutch oven. You wind up with an extra skillet to wash, but I feel the chicken browns better and the second batch isn't as likely to burn.

You can swap grocery store table grapes for the muscadines or scuppernongs, but you may need to reduce the sauce a little more. Remove the cooked chicken from the sauce, boil it down to the desired consistency, and then return the chicken to warm it through.

Concord Grape Pie

Serves 8

Pie is a labor of love. This pie especially. Halving and seeding nearly 3 pounds of grapes is messy and time consuming, but the results are outstanding. I first made this in 1997, and it haunted my memory for years. I was always on the lookout for Concord grapes after that, but rarely did I find them. I like to make this with a traditional piecrust, but the Rye Pastry on page 356 would be worth a shot.

2½ pounds (1.1 kg) stemmed
    Concord grapes
¾ cup (150 g) granulated sugar
⅓ cup (40 g) cornstarch
Pinch ground cloves
Pinch salt
Finely grated zest of ½ lemon
1 teaspoon lemon juice
Enough pie dough for a single-
    or double-crust pie
1 egg yolk
1 tablespoon heavy cream

NOTE: Dumping a hot filling into an unbaked pie shell may make the crust greasy and soft, and it may slump down the sides of the dish, so make sure it's somewhat cooled down before you put it in the lined pie dish.

One to one and a half hours is a wide range, but in my experience pies take as long as they take to bake. You're best off going by visual cues to judge doneness.

Halve the grapes. One by one, remove the seeds. (You can use the clean end of a large bobby pin, but if the grapes are soft, only your fingers will work. It's not hard, but it is slow going.) Put the seedless grape halves in a large strainer set over a large bowl.

In a medium saucepan, stir together the sugar, cornstarch, cloves, salt, and lemon zest. Add the lemon juice, strained grape juice, and half of the grape flesh; stir to combine. Over medium-high heat, bring to a boil, stirring constantly. Cook until thick, about 2 minutes. Remove from the heat, and stir in the remaining grape flesh. Scrape into the large bowl and chill.

Preheat the oven to 400°F (200°C), and position a rack in the center. Grease a deep 9-inch (23 cm) pie dish. On a floured surface, roll out the pie dough, and transfer it to the prepared dish. Trim the crust and crimp the edges decoratively, if you like. If you're using a top crust, roll that out (I like to cut out small circles using the wide end of a metal pastry tip to create a polka-dot effect).

In a small bowl, stir together the egg yolk and cream until combined.

When the filling is between tepid and cool, scrape it into the lined pie dish. Top with the top crust, if using, and pinch to seal. Cut a few vents in the top crust with a sharp knife for steam to escape. Brush the top crust with the egg wash.

Set on top of a sheet of foil over a rimmed cookie sheet or pizza pan to catch any filling that may bubble over. Bake until the filling is bubbling vigorously and the crust is nicely browned, 60 to 90 minutes. (Tent the pie loosely with foil if the crust starts getting dark but the filling needs to bake longer.) Set on a rack to cool. Serve warm or at room temperature. This is a very gushy pie.

Scuppernong Sorbet

Serves 6–8

The flavor of this sorbet is crazy intense, an explosion of grapiness. The yield here is small, with the intention of serving it in modest portions. You can double the recipe if you'd like more, but make sure your ice cream maker will accommodate all 6 cups (1.5 L) of the sorbet base in a doubled recipe.

2 pounds (910 g) scuppernong or muscadine grapes, quartered
⅔–¾ cup (130–150 g) granulated sugar
1½ tablespoons freshly squeezed lemon juice
½ cup (120 ml) water

In a large nonreactive bowl, toss together the grapes and sugar. Let them sit at room temperature for 30 minutes to macerate. The grapes should break down a little bit and release their juice.

Transfer the grapes and their juice to a conical food mill or fine-mesh strainer. Press down to release as much juice and goop as you can. Use a rubber spatula to scrape any goop still clinging to the exterior of the strainer. Add the lemon juice and water.

Transfer to an ice cream maker, and freeze according to the manufacturer's instructions. When the base is ready, it will look like a thick Slurpee. Transfer to a shallow metal pan, such as a standard loaf pan (this makes it easier to scoop later), cover with plastic wrap, and freeze until scoopable and set, about 2 hours. The sorbet will keep for about a week, but as it sits in the freezer it becomes more difficult to scoop. Let it soften for 5 to 10 minutes for maximum enjoyability.

NOTE: You can make this with muscadines, but the natural color of the grapes will lessen somewhat.

VIRGINIA CREEPER

Parthenocissus quinquefolia
Vitaceae family
Eastern and central US and Canada

This native vine climbs heartily on buildings, fences, and trees all over temperate areas of eastern and central North America. It is not very picky about the kind of soil it grows in. It's oftentimes confused with poison ivy, though their leaves are not very similar: Poison ivy has "leaves of three"; Virginia creeper has five leaves. In the late summer blue-black berries grow dangling from short red stems. They could be mistaken for a number of edible fruits, such as wild grapes or wild raisins (nannyberries). In fact, false grapes is one of its many colloquial names (it's in the grape family). Another is five leaves.

I often see Virginia creeper entwined with the vines and branches of grapes, wild raisins, and pokeweed. In the fall its leaves turn an attractive bright red. It's pretty recognizable, but its prevalence is a good reminder to slow down and look carefully before thinking about picking anything.

Like many poisonous-to-us plants, Virginia creeper is useful to animals. Deer and small mammals may eat its berries, but birds favor them more. It's also a good provider of habitat for insects and birds. Virginia creeper's ability to steadily creep up trees can eventually deprive them of sunlight and kill them. It is helpful and potentially sinister all at once, depending on where it grows.

GROUND CHERRIES

Physalis spp.
Solanaceae family
Throughout the US; eastern and central Canada

FOR YEARS I SCANNED THE "COOK'S CORNER" recipe exchange column in the Wednesday *Columbus Dispatch* food section. I was a kid, suspicious of all unusual foods, and I enjoyed the kooky quaintness of the older readers' requests. One summer a reader wrote in seeking recipes for ground cherry pie, as well as places to buy ground cherries. "Ground cherries!" I thought, fascinated. I imagined bizarre cherries that grew like fruity radishes in the ground.

Alas, ground cherries are not a succulently sweet root vegetable, but they are a wonderfully old-fashioned plant that was, and is, beloved for pie by a certain generation. Ground cherries also have nothing to do with actual cherries: They do not grow on trees, they don't look like cherries, they don't taste like cherries, and they are usually smaller than cherries. The fruit is delicious in its own right. The best ground cherries have a pineapple-strawberry thing going on, and when you bite into them their taut skin cracks open and their juicy flesh bursts forth—a true summer treat.

Ground cherries belong to the nightshade family (a quick peek at their eggplantlike leaves is a dead giveaway), with close relatives including tomatillos (*P. philadelphica*) and Chinese lantern (*P. alkekengi*). Like its cousins, the ground cherry produces fruit encased in a papery husk, but the husks aren't as fiercely sticky as a tomatillo's, or as showy as Chinese lanterns. In fact, you have to peer under the plant's foliage to see the fruits, which are shy little things. They knock tomatillos out of the ballpark, in my opinion, and they easily best Chinese lanterns, which are toxic.

Worldwide, the *Physalis* fan can find over seventy varieties, and plenty of them are ground cherries native to North America: the clammy ground cherry, the smooth ground cherry, the long-leaved ground cherry. Wild ground cherries tend to grow in well-drained areas, are apt to appear in fields tilled for agriculture, and are capable of wreaking a little havoc by hosting diseases and viruses of commercial crops. A number of states classify them as weeds—lucky you! But some types of ground cherries are more palatable than others, so don't get excited yet. Also, if you spy ground cherries near a tilled field, consider if there may be agricultural runoff before you sample any.

The marvel of ground cherries is their appearance in meadows and fields, spots where trees bearing familiar fruits like apples and pears were not conveniently already growing for homesteaders of the Westward Expansion. But those homesteaders could grow or forage for ground cherries: Harvesting and preserving the fruit is mentioned in Willa Cather's *My Ántonia*, as well as memoirs about homesteading in Oregon by Fannie Adams Copper and letters by Minnesota homesteader Mary E. Carpenter.

Cultivating ground cherries is another thing. Their seeds (which a mature plant produces in abundance, up to thirty thousand!) can be fussy to germinate. It's a smart idea to start them indoors at the same time you'd start tomato seeds. I grow mine in containers—it's one of the few fruits my teeny backyard will accommodate—and in a single season, the thing goes from a teeny seed to a sprawling plant that produces handfuls of adorable fruits in easily manageable waves. I delight in their lacy little hulls, the berries like golden pearls in a filigree setting.

One of the most delicious varieties, *P. peruviana*, hails from South America and is also called Cape gooseberry (though ground cherries are not gooseberries at all). In recent years the ground cherry has been grown on a larger commercial scale (well, large compared with next to nothing) and marketed as "goldenberries," a superfood. You may see them in fancier food markets, sold in plastic clamshells.

Ground cherries enjoy a bevy of folksy nicknames, like strawberry tomato or husk-tomato or husk cherry or bladder cherry. In Hawaii they're called *poha*, and they are loved as both plants and ingredient (particularly in jam). The plant was first recorded there in 1825 and now grows wild, though it's cultivated there, too.

BITTER NIGHTSHADE / EUROPEAN BITTERSWEET

Solanum dulcamara
Solanaceae family
Throughout the US and Canada

This perennial vine is as likely to show up in neglected flower beds as it is near a marsh or river. Its flowers are small, a lovely purple with a yellow center. Its berries ripen from a hard green to a brilliant red and resemble tiny cousins of tomatoes. They are ripe when tomatoes happen to be ripe: high summer to early fall. They're both in the nightshade family, but tomatoes are great to eat, and this guy isn't.

All parts of the plant are toxic. Supposedly it has its "bittersweet" name because the stem and roots, when chewed, taste bitter and then sweet, but I'm not going to put that to the test myself. Just handling this plant can give you a rash. The berries are alkaline (as nightshades are wont to be), and eating a small quantity can induce nausea, vomiting, and other unpleasantries.

Native to Europe, Africa, and Asia, bitter nightshade was introduced as a cultivated ornamental and got somewhat out of hand. It grows in North America from coast to coast, more in the North than the South. It is a bane of gardeners, as it can be tricky to eradicate once it settles in. I think it's cute, but then again, it's not growing in my yard.

Harvesting and Storage

In most places ground cherries will be ripe around July or August. Don't eat the leaves or unripe berries—they're not a stand-in for tomatillos, and they'll make you queasy. It's easy to tell when you're on the right track, because the husk of a ripe ground cherry turns yellow, and then the fruit usually falls to the ground. The husk will be a little crackly and dry. Pull it off like you are opening a present, and eat the berry inside.

If any of the fallen ground cherries you gather are not quite as luscious tasting as you'd like, take heart: Harvested ground cherries will continue to ripen at room temperature. They'll just get better and better. Ripe ground cherries will stay good under refrigeration for five days or longer, but I simply keep them out at room temperature and prey on the ones with the most dried-out husks (keep them in their husks until you're ready to use them). I've read that they'll keep spread on trays or in mesh sacks in cool, dark places for up to 3 months, but I've never had any stick around that long. I'd recommend freezing only in instances when you plan to cook or puree the fruit later; remove the husks first. For long-term storage, dehydrate ground cherries and render them into raisin-y nibbles (those seeking their nutritional benefits often buy them this way, as they are easier to store and transport).

Culinary Possibilities

I like the first one of the year best, eaten out of doors while it's still warm from the sun. Ground cherries are the perfect plant for people who like to indulge in garden snacks, or those who go on country walks in open areas between vast planted fields. But otherwise, ground cherries are tastiest after they've ripened a couple of days off the plant.

The bright and sunny flavors come through best when the fruit is enjoyed raw, or flash-cooked (despite the yearnings of that *Dispatch* reader so many years ago, pie is not my favorite way to put them to use).

Savory-sweet raw relishes and salsas dotted with fresh herbs show them off to fabulous effect. If you have a bumper crop to contend with, the usual bonanza of jams, chutneys, and preserves will do just fine (bonus: Ground cherries are high in pectin).

HACKBERRIES

Celtis occidentalis
Cannabaceae family
Eastern and central US and Canada

VARIOUS SPECIES OF HACKBERRY TREES grow all over the place in diverse contexts: wild backcountry, manicured city parks, the narrow strip of soil between sidewalks and buildings. They're easy to identify, plentiful, and often easy to access. But don't pop their pea-sized fruits in your mouth too casually, or you'll crack a molar. Hackberries are almost all seed, and those buggers are very hard.

Accordingly, hackberries will make more sense if you think of them as burly seeds instead of berries: not luscious and juicy, but densely packed bundles of energy-giving fuel and flavor. The rub is figuring out what to do with them—but wily foragers are often not easily discouraged, and your journey into hackberry lore will fulfill you in multiple ways.

Once you learn to zero in on a few key characteristics of the common hackberry tree (which ranges from eastern to central states), you'll be identifying them in your sleep. Once classified in the elm family, they are now part of the Cannabaceae family, which also includes hops and marijuana. Every now and then one of these reclassifications comes along to remind us that our understanding of the natural world evolves with the natural world itself, and sometimes our feelings take a while to catch up. Science downgrades Pluto to a dwarf planet, and we mourn the loss of a comforting notion about our grasp of cosmic order. The object itself remains the same, though; it's we that

have to make adjustments. In any case, hackberries do not have the kinds of properties that Willie Nelson and Snoop Dogg extol in song.

Their bark is unusual and immensely tactile. Often described as "warty," hackberry bark is deeply ridged, and if you look very close you'll notice distinct layers in those ridges—almost like the Grand Canyon. Another dead giveaway are bulbous protrusions from the undersides of the leaves. This is hackberry nipple gall, caused by psyllids, tiny insects that infest the leaves. It's common enough in the East that you'd be hard up to find a hackberry tree without it. Fortunately, it does not affect the health of tree or berries much.

The pea-sized berries are not at all showy. They dangle from long stems and slowly ripen from green to dull bluish black and finally to brown. In the late summer and fall, you'll want to keep closer tabs on them, and once they turn the color of milk chocolate, start sampling. Gently bite down on the almost crunchy shell of the berries, which will ideally offer a thin layer of grainy sweetness that calls to mind a less fleshy date, offering notes of honey and raisin. Quickly you'll realize why some species of hackberry found in the South and Southwest (*C. laevigata*) are called sugarberry. They are indeed somewhat reminiscent of candy.

Then you'll hit the seed, and perhaps all you can do is spit it out, because some of them can be rock-hard. The seed of every single hackberry I've sampled was too resilient to break into without risking a trip to the dentist, but since there's a precedent of people using entire crushed berries, I have not given up hope.

High in carbohydrates and essential vitamins and easy to store for long periods without spoiling, hackberries were likely a valued part of a varied diet for hunter-gatherers in North America. Paleo-Indian and Early Archaic archaeological sites in Texas, Mississippi, Alabama, and Pennsylvania show evidence of hackberries accounting for at least some portion of their diet. Pounding the hackberries to a pulp was the first step for preparations such as porridges, seasoning pastes and powders, and dense cakes.

Today, hackberries are common street trees because they grow in a wide range of soils and are resistant to drought. Walking down one block on the main drag of my town, I pass at least three, while along the nearby bike path I spot so many I lose count. They are as comfortable being city slickers as they are country folk.

Picking hackberries is fun if you're tall, or the tree is short. There are a few time-saving hackberry hacks, though coaxing young children to help is of little use here; likely they won't be able to reach the branches. Those confident in their abilities can climb the tree and try shaking the berries loose to fall on a tarp underneath. Whether you take the batch approach or do it one by one, waiting until the leaves wither or drop is helpful, because the berries will be much more visible. Thankfully, they'll cling to their branches for months.

If you don't want to wait for the seasons to turn and the weather to get crummy, you can do it hackberry by hackberry. Usually all the berries ripen on a tree according to the same schedule, but you might have to be selective. If you don't like the looks or taste of one tree's offerings, hit up another. It may take an hour to get just a quart or so. I get on my bike and pedal around from tree to tree, stripping accessible branches and moving on once my attention wanes or the picking gets tricky; mixing up the scene that way helps break up the monotony. Listening to podcasts in your headphones does, too.

So-called hackberry milk is the best entry point for hackberry experimentation. It's not so much a dairy-free milk alternative as it is a thin but tasty liquid. Make it by pureeing ripe berries with water, though the puree step is not as easy as it sounds. And you gotta get the berries first.

This is all to say hackberry milk is probably not going to be a staple of your diet unless you spend a few weeks every fall doing little but harvesting and storing hackberries. One reason to make it is to connect with a contemporary approximation of aboriginal foodways. Another is that hackberries, on the surface, don't look like they have much to offer, but now you are in on their secret.

Harvesting and Storage

Sturdy and low in moisture, hackberries are terrifically low-maintenance—no need to be gentle with these babies! I harvest them in an empty quart-sized (1 L) plastic yogurt container because it's easy to hold, lets me know when I have enough hackberries to make hackberry milk, and has a handy lid to prevent spills as I transport them home.

Hackberries will keep for months—no need to freeze. Lay the berries in shallow trays or put them in mesh sacks to allow good airflow, and you're set. They'll wrinkle up just a bit, but they won't dry like other fruits you're used to seeing.

As fall turns cooler, the berries remaining on the tree will get wrinkly the same way. You can continue harvesting them pre-dried in this manner all winter long. I've not noticed if drying this way makes their seeds any softer, but I'd like to think it's possible. Their pulverized fragments can be unappealingly gritty.

Culinary Possibilities

Hackberries are heavy duty, rich and sweet. The fastest and most convenient way to render many seeds edible in our time is to grab a high-powered blender. Pounding hackberries with a mortar and pestle—even a big, burly granite one—will quickly make you appreciate modern conveniences. I've seen recipes for hackberry jam, which doesn't seem to make much sense for a gritty fruit with barely any flesh to speak of. Go for it if you like, but keep in mind it is not written in stone that every foraged berry must be made into jam.

Slightly sweet hackberry milk is better as a beverage or ingredient than a cereal topper. Use it instead of coconut milk in stews and soups, and then consider how foraged fruit can show up in the unlikeliest of end destinations.

Hackberry Milk

Makes about 2 cups (480 ml)

Hackberries are incredibly low in moisture, so to render them into liquid, you must add water. What you wind up with is a "milk" that's somewhat sweet and chalky, with a lovely orange-ish shade, as if you tinted almond milk with turmeric. You can drink it cold or gently warmed. It's fun to experiment with in cooking applications, too—use it as a base for breakfast grain porridges, or add it to sauces and stews. Hackberry milk does not have the same rich body that canned coconut milk does, but its sweetness makes it well suited to curries or stews that excel with the interplay of tart, sweet, and salty tastes.

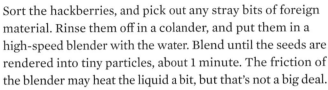

1 cup hackberries

2 cups (480 ml) cold water

Sort the hackberries, and pick out any stray bits of foreign material. Rinse them off in a colander, and put them in a high-speed blender with the water. Blend until the seeds are rendered into tiny particles, about 1 minute. The friction of the blender may heat the liquid a bit, but that's not a big deal.

Strain the liquid through a nut milk bag, jelly bag, or strainer lined with fine-weave cheesecloth. This could take a few hours. Wring the bag to squeeze out as much liquid as possible. Store in a jar, and refrigerate for at least a few days—I've never had it around for very long.

The hackberry milk will separate into two layers as it sits; there will be a clear upper layer and a silty tan layer underneath. A quick stir will bring it back together.

NOTE: Don't attempt this with a regular blender. Or do attempt it, but don't assume it will work. You can see this is a volume ratio of 1 part hackberries to 2 parts water, so scale up or down as you like—though for the sake of blender capacity I wouldn't more than double it.

The solids left after straining will be very gritty, with seed fragments that may be as hard as coarse sand. This is bad news for your teeth. After all the work of collecting hackberries and rendering them into a fascinating and unusual liquid, you may want to see if you can make use of this stuff, too. I tried drying it out in a low oven and then running it through an electric coffee grinder to make a flour, but it didn't alter the size or hardness of the particles.

The Forager's Palate

Lesser-known fruits can lure foragers to fields and forests because they promise untasted flavors.

They are unusual only because we are not used to experiencing them. The industrial agriculture and food production that feeds our world relies on monoculture—the Cavendish banana, for example, all but dominates the marketplace, so when we imagine a banana, we imagine the Cavendish. Fruit breeders create products that grow quickly, have a long shelf life, and look appealing. Taste is a consideration, but not *the* consideration. It's like music that's been run through an effects pedal, but instead of giving the music more texture, it erases its personality.

I've noticed that foraging enthusiasts tend to have sturdier palates than the average person—they relish bitter, pungent, and sour tastes that can alienate others. This does not mean foraging enthusiasts are superior tasters, but I do wonder if some of them are simply born with a more open palate; that is, maybe supertasters (those whose taste

buds have the most taste receptors) are put off by the typically more intense flavors of foraged foods.

The other thing at work here is cultural: Familiar flavors and textures are what we crave and return to. If a time machine transported an

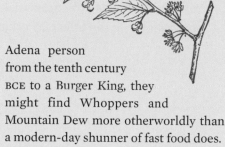

Adena person from the tenth century BCE to a Burger King, they might find Whoppers and Mountain Dew more otherworldly than a modern-day shunner of fast food does.

I used to love Whoppers when I was a teenager, but the last time I had one I thought it was gross. Either the Whopper changed or I did, but likely both. The more I expand my foraging repertoire, the more adventurous my palate becomes. It happens naturally, bit by bit; there's nothing forced about it, but as a chef, I'm sort of pre-selected to be that way. Chefs are always looking for ways to expand their vocabulary of flavors. Let's say you're not a chef or a nerdy plant person. If the wild taste of wild foods isn't to your liking, there's nothing wrong with sticking to some trailside raspberries. Taste is only one part of the foraging experience.

HUCKLEBERRIES

Vaccinium spp.
Ericaceae family
Throughout Canada; northern and western US

HUCKLEBERRIES ARE WILD through and through, and a certain type of person with a fierce independent streak and a love of self-sufficiency sees huckleberries as an emblem of a western way of life. Northwestern Montana is known for its huckleberries, as are Washington and Oregon. It's the state fruit of Idaho. Species grow all the way up the Pacific Coast to Alaska.

Everyone has heard of huckleberries, but relatively few actually get to taste them. They are true foragers' delights, and it is unwise to describe them as similar to blueberries around a huckleberry hound, because you will get an earful. Huckleberries have a more prominent "belly button" on their blossom ends than blueberries do. While blueberries can grow in clusters of several berries, huckleberries stud branches one by one, asking your fingers to be more nimble. And most important, huckleberries taste like huckleberries: intense, juicy, addictive.

The huckleberry hunt can get competitive, but there is a precedent for working things out. Huckleberries were at the heart of a treaty between the Yakima Nation and the US National Forest Service. The contested huckleberries were in Gifford Pinchot National Forest, about 100 miles (160 km) south of Seattle. The Yakima have foraged for huckleberries on the land for generations—to them, the annual picking and preserving of berries is a culturally, socially, and spiritually significant event—but during the Great Depression, outsiders began showing up and stripping the berries, too. In 1932 forest supervisor J. R. Burkhardt met with council members and eventually set aside 2,800 acres (1,135 ha) for tribal use during huckleberry season. The agreement was bound with a handshake and eventually written into the forest's management plan. It is still in effect today, though reputedly some non–Native Americans choose not to heed the signs posted and harvest huckleberries freely.

Speaking of national forests, huckleberries and hiking go hand in hand. While out on unrelated mountain or meadow adventures, you can scout out promising spots to return to. Serious pickers think nothing of going to higher elevations to get the best ones. The deeper the season, the higher you go.

Huckleberries are not big, and to pick many is a legitimate outing. Because huckleberries thrive on slopes, harvesters must be intrepid. There is brush to get through and footing to maintain. A fall could mean spilling your container, a small tragedy. Transfer your open container to a lidded container or zip-top bag every now and then, so if you do slip, your loss will be minimized.

The bushes can be low, or grow up to 6 or 7 feet (1.8 to 2.1 m) tall. Some people spot huckleberries on the side of a road, a sign that there are probably more huckleberries up higher. Pull over and check it out! If it's been cleared of berries already, there may be other bushes farther off the road to investigate.

Forest fires are part of what make huckleberries grow, though it's a slow cycle. Since huckleberries like full sun, they eventually establish themselves in open areas left after a burn. There's a push-pull of vegetation as the forest begins to encroach on its old territory, and Native Americans would some-

times burn trees and brush to preserve huckleberry fields.

Most species of huckleberries are dark purple-blue-black when ripe, but red huckleberries (*V. parvifolium*) are red when ripe, and rumored to be more tart than black ones. After one taste, you'll know!

Harvesting and Storage

Ripe berries are plump, sweet, and dark purple. They are soft, so pick them one at a time.

Pick out any debris once you get home, but don't rinse the berries until you use them. Some huckleberry fans don't rinse them at all, because it will wash off precious juice.

Zip-top bags don't provide much cushioning, but they do contain any leaking juices. If you want intact berries, bring a rigid container.

Freeze huckleberries within a day of harvesting them. They get softer as they sit, and if frozen they form a block and thaw into mush after releasing more liquid—still usable for some recipes, but not always the desired effect. Par-freeze the berries on a rimmed baking sheet before bagging them so you can pull out individual frozen berries and drop them directly onto rounds of griddling pancake batter.

Culinary Possibilities

Huckleberries release their juice easily, so use them in ways that take advantage of this. Gooey jams and gushy baked desserts are obvious here. Anyone who's browsed in a Rocky Mountain or Pacific Northwest gift shop has doubtlessly seen plastic bears filled with huckleberry-infused honey. Quick-cooked ketchups and sauces for game bring it to the savory side.

Buckwheat Huckleberry Buckle

Serves 8–10

We reserve buckwheat for pancakes and blinis, which is a shame. Buckwheat flour has a strong but sophisticated flavor—a little nutty, a little fruity. It goes well with berries in this buckle—a rustic fruit dessert so named because the berries make the cake buckle in its center. Think of it as a giant muffin. I love this at breakfast, or as a midday snack.

For the streusel

½ cup (50 g) rolled oats
¼ cup (45 g) buckwheat flour
½ cup (110 g) packed light
    brown sugar
¼ teaspoon salt
Pinch freshly grated nutmeg
¼ cup (½ stick) unsalted butter,
    cut into small pieces and at
    room temperature

For the cake

1¼ cups (160 g) unbleached
    all-purpose flour
½ cup (90 g) buckwheat flour
1 teaspoon baking powder
¼ teaspoon baking soda
¼ teaspoon table salt
6 tablespoons unsalted butter,
    at room temperature
¾ cup (150 g) granulated sugar
2 large eggs
⅔ cup (160 ml) buttermilk
3 cups (435 g) huckleberries

Preheat the oven to 350°F (175°C) and set the rack in the middle. Grease a 9 × 9-inch (23 × 23 cm) pan.

Make the streusel: Combine the oats, buckwheat, brown sugar, salt, nutmeg, and butter in a medium bowl. With your fingertips, work the mixture until it's crumbly. Set aside.

Make the cake: In a medium bowl, whisk together the ingredients from flour through the salt.

In the bowl of an electric mixer fitted with the paddle attachment, beat the butter and granulated sugar on medium-high speed until lightened, about 2 minutes. Add the eggs one at a time, scraping down the bowl after each, and beat until fluffy, about 3 minutes. Mix in the flour mixture in three additions, alternating with the buttermilk. Fold in half of the huckleberries.

Spread the batter in the prepared pan. Scatter the remaining berries over the top, then the streusel. Bake until a toothpick inserted in the center comes out clean, about 40 minutes. Cool on a rack for at least 10 minutes before serving. Covered with plastic wrap, the buckle will keep for up to 2 days.

ALSO TRY WITH: Blueberries, mulberries, blackberries, raspberries.

Pan Swaps

Does a recipe call for a 2-quart casserole and you have no idea how much your casserole holds? Or is your only 13 × 9-inch baking dish in the refrigerator with leftover enchiladas and you don't feel like washing it?

Not all hope is lost! The easiest way to measure the capacity of a baking vessel is to fill it with water, and then measure the water. Meanwhile, if you don't have a baking dish of the correct dimensions, you may be able to use a dish with the same capacity. Be prepared for different results: longer or shorter baking times, gooey centers and overdone edges, or a cake that's thin and doesn't have as tender of a crumb but offers more surface area for smearing with delicious frosting. Sometimes a pan swap can create a result you like better, and sometimes it's a flop.

Also, if you do that math, a 9 × 9-inch pan and an 8 × 8-inch pan have very different capacities, even though they seem like they'd be pretty close. Your calculator can be useful for educated risk taking.

Talking to Plants

Plants are animate objects, and even if they can't hear what we tell them, I fully encourage you to talk to them. It's a sign of a happy and confident person who has no problem looking like a loon.

Usually I talk to plants just in passing, like when I'm running on my favorite trails in the woods and I glimpse mushrooms, or spot the first spring leaves of trees I've been missing all winter. "Hey, cutie!" or "Hi, friend!" I'll call as I go by. Someday, another person will be on the trail and wonder why I am greeting them so warmly. But I can't help it. It's a knee-jerk thing, just an impromptu prayer of thanks. There's something about externalizing and vocalizing your appreciation that I think gets better absorbed on a cosmic level.

I also kiss plants, if I'm feeling more subdued. Not make out with them, just grace their fruit or blossoms with my lips, or kiss my hand and then touch a tree trunk, like a pope of the forest. (Obviously don't kiss *unfamiliar* plants. You don't want any weird rashes.)

Language sets us apart as humans, and we are so gung ho about it we speak aggressively to nonhumans all the time: pets, traffic signals, television screens with sporting events playing on them. None of those things are going to talk back, but plants have a lot to tell us if you pay close attention over a span of seasons.

JUNIPER

Juniperus spp.
Cupressaceae family
Throughout the US and Canada

INCLUDING JUNIPER IN THIS BOOK is a bit of a cheat, since juniper berries—best known for flavoring gin—are not berries. The "fruits" are actually cones, and only the female trees bear them. Juniper is the one edible member of the cypress family, but not all junipers are edible. The three discussed here are.

People love or hate gin's piney, sharp flavor, and even those who love gin wouldn't love gin in every cocktail—it's too darn distinctive. Likewise, juniper berries are not the sort of berries one pigs out on, though they taste better on their own than you might imagine.

Juniper trees or shrubs like rocky soil. Depending on the variety or the time of year and their age, the berries are dark or light blue, and usually with a powdery bloom.

Common juniper (*J. communis*) can be found all across North America, but as it favors dry, rocky soil, it's not as easy to run across in states without that kind of landscape. These are the whole, dried juniper berries we see dried and sold in the spice aisle.

Eastern red cedar (*J. virginiana*) is a much more likely culprit for those who live in central and eastern regions of the United States. It's a smaller berry than common juniper, and it's also milder. We had two of these trees in my yard when I was a kid, and I recall not liking them because I thought the berries were ugly and the greenery was too

sharp and needle-y. The trees can grow huge (our backyard ones were only 20 feet/6 m high or so). Eastern red cedar was once the go-to wood of the US pencil industry.

California juniper (*J. californica*), true to its name, is mostly found in California. Indigenous people there used its berries fresh and dried as a food source, often in times of scarcity; they favored other parts of the tree for medicinal uses. You're just as likely to see California juniper growing wild in Joshua Tree National Park as you are to spot it as a drought-tolerant landscaping feature by a strip mall.

Junipers can be easy to mix up, because to the untrained eye, a lot of them look the same. How do you know if the juniper you've found is edible or not? If you're sure it's a juniper tree, period, then taste a berry. If it's not an edible variety, you'll know, because it will be unpalatably bitter. If it's sweet and compelling, it's edible.

Whole juniper berries are often added to recipes in teaspoons or tablespoons, but if you like eating them as is, don't go overboard. They have medicinal properties, and a nibble or two is sufficient. They can have a diuretic or laxative effect, and pregnant women should avoid them.

Harvesting and Storage

The berries start out greenish white in the spring and become darker bluish purple as they age. Look for the more mature ones with eastern

red cedar. The smaller berries are said to have a better taste. Some foragers prefer the flavor of green California juniper berries. In either case, this forageable requires a fair share of tasting and observing over the seasons, meaning certain trees will become touchstones to you throughout the year as you first become acquainted.

Pluck off the berries and refrigerate them in a sealed jar for 2 weeks. Freeze or dehydrate for long-term storage. Keep juniper berries whole until you intend to use them, as their volatile oils quickly fade when they come into contact with air.

Culinary Possibilities

Juniper berries are an ace in rubs for meats. Crush fresh ones to make a wet rub, and grind dry ones to add to dry rubs. Pancetta is cured with juniper berries, and stronger meats like lamb, beef, or game (venison! duck!) have the backbone necessary to be a worthy partner to juniper's assertive character.

In central and northern Europe, juniper is a common flavoring in rich or salty food. Slip a few berries into a batch of fermenting kraut, or add a couple to finished kraut as it cooks (juniper is a must-have for the sachet of herbs added to *choucroute garni*, the bonanza of pork and

Culinary Foragers

The very people most responsible for bringing foraged ingredients to a larger audience may not do much foraging themselves. Chefs tend to be driven, curious, and restless. Unfamiliar ingredients keep them on their toes. Ideally, spicebush or cattails don't show up on a menu because they sound far-out and novel, but because they push the limits of our flavor lexicon.

Running a restaurant isn't just about selecting the best baskets of morels and fiddleheads at the farmers market, though. There are constantly urgent issues to resolve: staffing dramas, broken dishwashers, unexpected reservation requests. Heading out to collect barberries at dawn isn't always possible.

Besides, some chefs have no burning urge to do every bit of the actual foraging. "I'm a cook, and I feel close to nature when I cook," the Danish-Italian chef Christian Puglisi said. Why shouldn't he?

The profit margin in food service can be disappointingly slender. In that mindset food is a product. But thinking artistically, food is a tangible result of plant life doing its miraculous work. Chefs have to consider both approaches at once. They may not have time to spend researching where foraged ingredients might fall in the larger ecosystem, and this is where things can get bumpy. Take ramps, for instance. The wild alliums are very slow to grow; it can take a single ramp up to four years to reach maturity. The chef-driven trendiness of ramps has led to unsustainable overharvesting. It can take a ramp patch a decade to fully recover, but few ramp-using chefs are aware of this. They just know that ramps are delicious, and that customers love to see them on menus.

The term *wild* always catches our attention. Wild figs, wild plums, wild mint. For people who are disinclined to tromp around in the backcountry—or even down the sidewalk—ordering a dish with foraged ingredients gives them an interface with nature. In most cases, though, the dish's remaining ingredients were not gathered by a fleet of earthy, buckskin-clad hill people. There's a big space between wildcrafted currant vinegar and a case of industrially produced red wine vinegar. Restaurants rely on both, and usually more of the latter, and there's nothing wrong with that. The thing that drives me batty is assuming "foraged" automatically means "good." Foraging ingredients isn't necessarily better for the Earth, or better for you. It's simply a way to obtain food.

The current mania for foraged items means that diners and chefs are hungry for connections, and foraged ingredients can spark conversations about where our food comes from and what it does for us—all of it, whether it's a CSA box or a field of mountain huckleberries or a shipment from Sysco.

LOOK BUT DON'T EAT!

YEW

Taxus spp.
Taxaceae family
Throughout the US and Canada

As a kid, never once was I tempted to try one of the tiny red berries that grew on the yew bushes in our front yard. It just didn't occur to me, and it's a good thing: The seeds are highly poisonous.

Yews are conifers, and like juniper, the lady plants bear berrylike cones. Each of the tiny red "berries" is an aril surrounding a single, large black seed. I've squished yew berries between my fingers a few times and sneaked a fleeting lick of the seed-free pulp, which is safe to eat. The flavor's not objectionable—sweet and vaguely fruity, with hints of raspberry and cherry—but certainly nothing worth risking one's own well-being for, should a seed sneak down your gullet somehow.

What I found most notable wasn't their taste, but the slipperiness of the smashed fruit, which reminded me of Astroglide. Even so, don't try making DIY sexual lubricants out of yew berries. Just start noticing them on the branches among the blunt, pliable green needles so ubiquitous in countless suburban landscapes and consider this plant's little-known dark side.

sauerkraut that's the traditional dish of Alsace). Add them to vinegar-based brines for pickles. Commercial breweries such as Rogue, Dogfish Head, and even Samuel Adams have brewed beers with juniper berries, and home brewers are following suit, experimenting with foraged juniper.

Fresh juniper berries can even have a place in the pastry kitchen. Process the berries with sugar to make juniper-infused sugar, or add small whole fresh berries to buttery shortbreads or scones.

If you're accustomed to common juniper, you'll need to tinker around with foraged ones to see how many berries you need to get the flavor you want.

LEMONS

Citrus limon
Rutaceae family
Southern US in zones 9–11

HOW WONDERFUL IT IS TO LIVE in a world where a lemon is never just a lemon! All of the citrus we are familiar with descended from four species: the pomelo, the mandarin, the papeda (a forerunner of limes), and lastly the citron, thick of skin and scant of flesh. Citrons are the ancestors of lemons.

Old songs and sayings about lemons never flatter the fruit. This is perhaps because lemons are not enjoyable straight from the branch, but it's a very shortsighted point of view. What is a perfectly cooked cod fillet without its final redeeming squirt of lemon juice? What is a bowl of salad greens, olive oil, and salt without the vital kiss of acid?

Foraged lemons are fantastic, because (assuming they are free of pesticides) you can use the peels and zest without fear. (Any pesticides applied to citrus are more concentrated in the peels, which is why some recipes calling for a lot of citrus zest specify using organic citrus.) Those who live in places that are sunny year-round will do well to seek out lemon trees, which will flower and fruit sporadically all year, though in climates that are *quite* warm the spring and fall will see bigger loads of fruit.

Natural mutation, cross-pollination, and human manipulation have led to the lemons we widely buy in stores today: Eurekas, which are slightly elongated with dimpled skins, and Lisbons, which have more

prominent nipples and smoother skins. There's an exciting Variegated Pink Fleshed Eureka, striped green and yellow when ripe and holding pink-tinged flesh inside, though good luck running across one of those.

Happening upon a Meyer lemon tree is much more likely. In the early 2000s Meyers crossed over in public opinion from being something neat to grow to being something neat to eat, and this onetime novelty fruit is now available at mainstream grocery stores. Its name carries cachet in epicurean circles, but its increased presence in recipes and on menus hasn't dampened its allure to me.

Ruth Reichl, onetime editor of *Gourmet*, stated in an interview that Meyer lemons are overrated. "I am not a Meyer lemon person . . . I like acid. They're muted. I hate the fact that everyone loves Meyer lemons." Now I think of Ruth Reichl every time I see them or buy Meyer lemons, and I buy them every time I see them. Instead of thinking of Meyer lemons as lemons with no spine, I see them as their own specific citrus. They are mild, with an enticing floral character. I just can't get enough of them.

I tasted my first Meyer lemon not long after moving to California's wine country in early 2000. I journeyed there with dreams of becoming a wine writer, but instead became a pop music critic, an unpaid yet undeniably cooler hobby I subsidized with a job shelving books at the central branch of the county public library.

An anonymous fellow library employee had a tree in her yard, and she'd bring big paper grocery sacks straining under the weight of the Meyer lemons they held. I brought those sacks home, because they were free, and made their contents into lemonade using an extremely complicated procedure I'd read about in *Cook's Illustrated*.

I had no idea the lemons I was mashing up were Meyer lemons; I just thought they were weird but neat. They had a slight herbal smell, and their pretty yellow-orange peels were thinner than what I was accustomed to. They perplexed and delighted me. I took to mixing the resulting fragrant lemonade with Hamm's to make shandy, something I liked drinking after the long runs I'd take on gravel roads crisscrossing Sonoma's rolling stands of madrone and eucalyptus trees. I was twenty-three and had very few obligations in life. Everything lay ahead of me, little pockets of promise to hunt like Easter eggs.

I later quit my library job and eventually left California altogether, ditching the land of free hybrid citrus and hills stretched on the horizon like big, golden sleeping dogs. Now, a kid and a husband and a dozen jobs later, just handling a Meyer lemon triggers endorphins in my brain. I equate their flowery perfume with potential, with freedom, with can after golden can of Hamm's poured into a sweaty glass half filled with the most exquisite lemonade I'll ever make. I don't blame Ruth Reichl for getting peeved with the fuss people make over Meyer lemons, but I'll never pass up an opportunity to go past-tripping when I spot an overpriced bag of these rare beauties in my Midwest grocery store.

Meyer lemons got their name from Frank Meyer, a US Department of Agriculture employee whose job was to import new plants to the country. Meyer (née Frans Nicholas Meijer) was an émigré from the Netherlands. An official USDA certificate he carried on one of his expeditions stated he was traveling with "the purpose of Aiding in Agricultural Development, especially along the line of Pomology." Between 1905 and 1918, he took multiple expeditions through Asia to

collect plants. A 1908 photo of Meyer shows him in rugged explorer mode, standing with a walking stick against a backdrop of boulders. On the ground to his left are several burlap sacks of plant specimens.

Meyer introduced over twenty-five hundred plants to the United States—not just the Meyer lemon. But his life came to an abrupt end when he fell overboard from a passenger boat in the Yangtze River in 1918 and drowned. The oft-repeated origin of Meyer lemons begins with Meyer's "discovery" of this fruit in exotic China, but obviously they existed before an employee of the US government came traipsing around, and that's the part of the story I like to ponder. Some scientists speculate Meyer lemons are a hybrid of a true lemon and a mandarin orange, sweet orange, or bitter orange. Others consider it to be a selected cultivar of the lemon, and refer to it as *Citrus limon* 'Meyeri'.

In outdated literature about citrus cultivation, you'll sometimes see Meyers vilified as a "scourge." That's because Meyer lemon trees were a carrier of the citrus tristeza virus, which could spread to other citrus trees—not just Meyers. Happily, the current Meyer lemon variety (Improved Meyer, introduced in 1971) is resistant to the virus.

The trees are compact, more cold hardy than other citrus varieties, and don't require much pruning—all reasons they are a popular choice for landscaping. The peak of fruit production is late winter through early spring. If you don't grow one, the next best thing is to live close to someone who does.

For those who don't reside in Mediterranean-esque climates, growing Meyer lemons indoors in containers is the next best thing (in fact, they are one of the easiest citrus varieties to grow in containers). While your annual crop will be comparatively minuscule, the rewards are a cheerful sight to have inside in the winter—especially if you're savoring your petite, sunny harvest when there's still snow on the ground outside.

Harvesting and Storage

Look for lemons with skin that's supple, shiny, and yellow. Leave lemons with green skin on the tree to ripen more. Lemons a few days past peak ripeness will taste flat and won't be as juicy.

Every restaurant I've worked in had a plastic bin full of lemons in the walk-in, and inevitably one on the bottom would go bad and set off a white mold chain reaction, spoiling the lot. Lesson: Do not store lemons all piled up. Do refrigerate them, which will keep them from getting wrinkly and tired looking. Kept in a sealed plastic bag, they'll stay pert for a month or so.

Thin-skinned Meyer lemons need slightly more careful handling. I've refrigerated them for up to a month, but they are quicker to mold than other citrus. Use them within a week or so to be on the safe side.

Culinary Possibilities

Look in the test kitchen of a big-time food publication, and likely you'll see a pile of zested lemons kicking around in a crisper drawer. Lemon zest adds that bright citrus flavor without pushing the acid level out of balance. Zest lemons with abandon, adding them to vegetables (brassicas like broccoli and cauliflower love the lift), rubs for lamb or veal, and fruit desserts galore.

Unless you run a restaurant, a single productive lemon tree could provide more lemons than even a sane cook can burn through. Fortunately, our twin crystalline vices, salt and sugar, transform the fruit when applied liberally. Cure whole lemons in salt and eventually you get an insane lemon drop smell, though sometimes it reminds me more of Pledge furniture polish. Preserved lemons last forever and ever and taste like lemon zest crossed with capers. I dice the peels finely and add them to things I'd normally add lemon zest and capers to—a lot, really.

As for sugar, there's candied lemon peel, which is quite good when you make it yourself (sorry, grocery store citron tubs). Custardy lemon curd freezes well to fill tart shells and swirl into ice cream in the future. And as the old saying goes, you can always make lemonade. If you are then still awash in lemons, consider beauty treatments.

LEMON BARS

Makes 16 large or 32 small bars

Lemon bars always have too much crust and not enough filling for my liking. I fixed that by scaling up the filling for my favorite lemon bar recipe by 2.5, and now *this* is my favorite lemon bar recipe.

For the crust
2 cups (260 g) unbleached
    all-purpose flour
½ cup (100 g) granulated sugar
½ teaspoon table salt
¾ cup (1½ sticks) cold butter, cut into
    ½-inch (1.3 cm) cubes

For the filling
6 tablespoons unbleached
    all-purpose flour
1½ cups (300 g) granulated sugar
7 large eggs
6 tablespoons heavy cream
2½ teaspoons finely grated
    lemon zest
¾ cup (180 ml) fresh lemon juice
¼ teaspoon table salt
Confectioner's sugar, for dusting

Preheat the oven to 350°F (175°C). Line a 13 × 9-inch (33 × 23 cm) baking pan with parchment paper, leaving an overhang of several inches along the long sides.

To make the crust, pulse the flour, sugar, and salt in the bowl of a food processor until combined. Add the butter cubes, and pulse until the mixture resembles coarse cornmeal. Press evenly into the prepared pan. Bake until golden brown, 20 to 25 minutes.

Meanwhile, make the filling. In a large bowl, whisk together the flour and sugar until combined. Whisk in the eggs, then the cream, zest, juice, and salt until well combined. Scrape over the hot baked crust, and bake until set, about 30 minutes. Set on a wire rack to cool completely.

Refrigerate the bars in the pan until cold (this makes them easier to cut). To cut the bars, place a large cutting board over the top of the baking pan. Invert, remove the pan, and peel off the parchment paper. Set another cutting board gently over the bars and invert again so the bars are filling side up. Cut into the bars with a long, sharp knife.

Before serving, dust the bars with powdered sugar. They will keep, tightly covered in a container and separated with waxed paper between layers, for 3 days.

ALSO TRY WITH: Sometimes I use a mix of citrus: lemons and limes, or lemons and oranges. I even used a grapefruit in there once.

AMAZING MEYER LEMON LEMONADE

Makes about 5 cups (1.2 L)

Here it is, the lemonade that sold me on Meyers. I read about this technique years ago in *Cook's Illustrated*, and as I recall, they'd discovered it in a nineteenth-century cookbook. The details are clouded in the haze of memory—the important thing is the method itself, which yields an intensely flavored, sophisticated lemonade because of the citrus oils released during the pummeling step. Also: Mashing lemons is wonderful for anger management.

If you have many lemons—Meyer or not—you can scale this recipe up. It's more about the ratio and method than exact measurements.

1 pound (455 g) Meyer lemons

Up to 1½ cups (300 g) granulated sugar, divided

4 cups (960 ml) water

If the lemons are not organic, rinse them off and rub them dry. Cut them in half pole to pole, and then slice them crosswise into half-moons between ⅛ and ¼ inch (3 to 6 mm) thick. Accuracy and looks aren't important here, but the thinner the slices, the quicker their juice will release.

Dump the lemon slices into a deep, nonreactive stockpot or Dutch oven, and add 1 cup (200 g) sugar. With a potato masher, pound the slices until the sugar begins to dissolve and the lemons give up their juice. You can take a break for a few minutes, let the maceration do its work, and come back to pound some more. In the end, you'll have pummeled the lemon slices in a syrupy liquid.

Scrape the mixture into a wire-mesh strainer set over a medium bowl. Press with a sturdy wooden spoon to get the lemons to give up as much liquid as possible (it will probably be about a cup). Transfer to a pitcher, and add the water. Taste, and add more sugar, if needed.

Serve chilled over ice. I really like this with a little vodka or moonshine, but obviously virgin lemonade is exquisite, too. For a shandy, combine about 1 part lemonade to 1 part lager or ale (make sure it's not too hoppy, though).

Preserved Lemons

Makes one 1-quart (1 L) jar

If you've never preserved lemons before, you're in for a treat. Very few home fermentation projects require so little effort for such dramatic results. Preserved lemons have a smell that's incredible, as intense as lemon hard candies. Their flavor is that of capers married with lemon zest; it's not sour as much as briny. Anyone with a lemon tree (Meyer or otherwise) will benefit from this preparation. Preserved lemons keep in the refrigerator for ages, and a jar of them makes a wonderful gift.

10–12 lemons (about 1¼ pounds/570 g)
½–1 cup (130–260 g) kosher salt or coarse sea salt
2 teaspoons dried red chili flakes, optional

Rinse off the lemons, and dry them with a clean tea towel. Set aside three of the lemons.

One by one, cut the remaining lemons almost into quarters by placing them stem side down on a cutting board and cutting three-quarters of the way down, so the lemons are still intact at the stem end but the cut parts open up like a flower.

In a medium bowl, toss together the salt and red chili flakes. Pack the cavity of each cut lemon with salt, using up all of the mixture. Cram the salt-packed lemons into the jar and press down with your fingers or a wooden spoon to make them fit—you want them to be really jammed in there. Juice the remaining three lemons and pour the juice into the jar.

Let the jar sit at room temperature for a few hours. If there's not enough liquid to fully cover the lemons, juice a few more lemons and add the liquid until the lemons are fully covered. (A good way to keep the lemons submerged is to nestle a sterilized rock on top.) Leave at least ¼ inch (6 mm) headspace between the lemons and the top of the jar.

Seal the jar, and let it sit at room temperature in a dark place for at least 3 weeks and up to 3 months. Check on it from time to time; you'll notice the aroma shifting from fresh and juicy to something more concentrated and hyper-lemony, but it shouldn't get funky-smelling like, say, sauerkraut. You can swish the jar around to make sure the liquid (which is now a brine) continues to disperse and dissolve the salt.

Once the lemons are cured, they will keep in the refrigerator indefinitely. I mean it! I have a jar that's 3 years old, and those lemons are showing no signs of wear and tear.

NOTE: What do you do with these guys once they're cured? Preserved lemons are at once citrusy and pickle-y, and they work well as a foil for heavy meats and fatty sauces. Many Middle Eastern tagines call for preserved lemon, but don't feel obligated to stop there. Mince a little of those preserved lemon peels, and add to tartar sauce, aioli, vinaigrette, or coleslaw. Slice them into paper-thin shreds, and use them as a garnish for poached or grilled fish, or add them to dress up a tuna or chicken salad. Anytime you might consider adding lemon zest to a savory dish, preserved lemon is probably a good fit, too. But a word to the wise: Make sure to chop preserved lemon finely. Pieces that are larger than a pea will overpower a dish, as well as be texturally distracting. You can even use the pulp, as Moroccan-born chef Mourad Lahlou does; he rubs it on chickens to season them before roasting them.

CITRUS CURD

Makes 1⅔ cups (400 ml)

It takes time to make a fruit curd, but it can feel meditative to stand at the stove and stir. I adapted this from a story Lynne Curry wrote for the *Oregonian*; it's been my go-to lemon curd ever since. Sometimes I have other citrus on hand and do a mix. I love this on biscuits and scones. Try folding a big dab of it into a Summer Berry Fool (page 322), leaving sunny yellow streaks visible.

2 large eggs

2 large egg yolks

¾ cup (150 g) granulated sugar

½ cup (120 ml) freshly squeezed and strained citrus juice (see note)

2 teaspoons finely grated lemon, orange, or lime zest

6 tablespoons butter

In a medium stainless-steel, nonstick, or enameled saucepan, beat the eggs, yolks, and sugar with a whisk until the sugar is mostly dissolved. Whisk in the lemon juice and zest.

Place the pan over medium heat and cook, stirring constantly with a heat-resistant spatula, making sure to scrape the bottom and corners of the pan. Slowly, the mixture will turn more opaque and the spatula will start to make visible swaths through it, 10 to 15 minutes. Keep stirring until the curd is as thick as sour cream and coats the spatula, 2 to 3 minutes more.

Remove the pan from the heat, and stir in the butter until it melts and the curd is smooth. Strain the curd through a fine-mesh sieve into a medium bowl, and lay a piece of plastic wrap directly on the surface to prevent a film from forming. Chill in the refrigerator until cool and set, at least 4 hours.

Transfer to a clean glass jar, and cover tightly. The curd will keep for about a month in the refrigerator, or in the freezer for up to 1 year.

NOTE: I've used all limes in this, all lemons, and a mix of lemons, oranges, and grapefruit. Grapefruit alone is overpowering, though.

Fruit Envy

There is probably plenty of fruit in your neighborhood to keep you busy, if you look hard enough. It might not be the fruit you want, though. The list of fruits I yearn for but generally can't find here is long and wistful: figs, huckleberries, cactus pears, any kind of citrus. And quince. I am constantly on the prowl for quince trees.

I see friends in faraway places sharing photos of their local fruits—papayas, in one case!—and I have to content myself with what I have here, which is a lot. Sometimes it's not the easiest to gather or deal with, but accepting what I have instead of what I wish for is a lot easier than buying a plane ticket just so I can rent a canoe to gather mayhaws, or driving across the country to luxuriate in bushels of pomegranates. That's when foraging mutates from an inexpensive pursuit to an extravagant lifestyle.

Fruit envy is as old as fruit itself, and there's nothing wrong with it inspiring road trips or vacations. But since foraging is all about routine, don't let your excitement about what you can't get cramp your hunger for what you might find. Fruit envy can cloud your ability to see what's already growing down the street—even if it's nothing more than a handful of minute wild carrot fruits.

LEMON HERB SALT

Makes about ½ cup (120 ml)

Conventional citrus is sprayed with pesticides. This isn't breaking news, but a lot of it is more concentrated on the surface—the peel. If you have lemons from a neighborhood tree, likely they're free of chemicals, and much better suited for using in recipes that call for tons of zest.

Mincing a pile of herbs and citrus zest into bitty flecks is (a) cathartic, and (b) easier than mincing a pile of onions. If you grow herbs and can't keep up with them, this is a perfect way to put them to good use in a manner that's shelf-stable but full of flavor. Give jars to your friends who don't eat jam. It's dead simple and well suited for cooks who prefer to eyeball measurements.

4 large sprigs rosemary

12 large sprigs thyme

Handful of fresh mint leaves

3 cloves garlic, minced

4–5 lemons

2 tablespoons coarse salt

On a large cutting board, stem the herbs and mince them, then mince the garlic. Zest the lemons over the whole works with a Microplane grater. Run your knife over the zest, herbs, and garlic again and again until it's all rendered into tiny, confetti-like bits. Transfer to a pie plate or rimmed dish, and stir in the salt with your fingers.

Let the plate sit out at room temperature for a few days until the herbs and zest are dry and flaky. Transfer to a clean, tightly sealed jar. For optimal flavor, use within 3 months.

NOTE: Use this as a finishing salt for meat, or sprinkle it over vegetables before roasting them. I love it on broccoli.

ALSO TRY WITH: Swap an orange for a few of the lemons. Or use a 50-50 mix of chives and parsley instead of the other herbs.

LIMES

Citrus spp.
Rutaceae family
Southern US in zones 8–11

LIMES ARE THE MOST ACIDIC of the citruses, and their usefulness to us generally hinges on the presence of other ingredients. A little lime goes a long way, and when it's missing, you know. What is a gin and tonic with no lime? A drink for desperate people, that's what.

Limes are synonymous with preventing scurvy on long sea voyages. Scurvy occurs because of a lack of vitamin C in the diet, and any number of fruits and vegetables can provide it. Why limes, then? Light and heat break down vitamin C, but citrus has keeping quality, and those thick rinds protect the essential nutrient. In 1747 Scottish physician James Lind established through clinical experiments that citrus cures scurvy. Vitamin C itself was not yet isolated—that didn't happen until 1932—so it took a long time for citrus to become a routine presence on ships. The Royal Navy didn't require it on its ships in foreign service until 1799. The citruses used were not limes exclusively, but lemons or whatever was available. Nevertheless, the daily dose inspired those in former British colonies to dub British sailors or immigrants from England "limeys."

Limes like it hot. Here is a good rule of thumb: If a location could be a logical setting for a Jimmy Buffett song, then limes will grow there. No Buffett, no limes. Though limes were introduced to Europe before they were to the New World, they didn't take hold there as much as lemons and oranges. Once Spanish explorers brought limes to the Caribbean,

they became naturalized (visitors observed them growing in Haiti in 1520) and eventually became an important ingredient in the region's own cuisine.

Key limes (*C. aurantiifolia*), also called Mexican limes and West Indian limes, are as close to the original lime as we're gonna get. Their size (small), complexion (blotchy), and color when ripe (yellow with green undertones) can confuse those accustomed to the so-called Persian limes that rule our produce aisles. Defenders of the Key lime say its flavor is both more aggressive and more complex, and that the iconic Key lime pie is not authentic without them. Detractors of the Key lime say it is an overrated pain in the ass. After grappling with a pound of firm, round little Key limes for far too long just to get a cup or so of juice, I can see their point, but Key limes are more perfumed, muskier, and generally more interesting than Persian limes. I grab Key limes up every time I run across them, which is unfortunately not directly from the trees. Mine come from well-stocked produce markets and are imported from Mexico.

It's interesting that we call them Key limes, because of all the places they grow, they have been in the Keys the shortest time. Their precise date of introduction is not known, but they were a cottage export industry for a short time in the 1910s and '20s, before they became the Keys' fruit mascot. According to *Fruits of Warm Climates*, "The fruits were pickled in saltwater and shipped to Boston where they were a popular snack for school children." I have a hard time visualizing this, but whatever. Key limes and the Keys suit each other. Both are offbeat, tiny, and not for everyone but very much for some people.

Key limes are true to seed and fun for doting types to grow indoors or out (given a Jimmy Buffett–appropriate climate or sunny and toasty corner indoors). There is barely a commercial crop in the United States, so the Key limes one sees will most likely be ornamental.

Persian, Tahitian, or Bearss limes (*C. latifolia*) are bigger, are oblong, and more readily give up their juice, though that's more a merit of their size than of their moisture content. They are the limes we think of when we think of limes. Finding either will set you up well to tie together a recipe featuring any combination of cilantro, mint, chilies, garlic, ginger, seafood, pork, chicken, and stinky fermented fish sauce or shrimp paste.

Harvesting and Storage

Limes ripen only on the tree, and not all at once. Key limes will be mostly yellow, and possibly with some brown (but not rotten) spots. Persian limes will be bright, deep green.

Lime trees bear fruit all year long. Older texts mention barrels for storage—perhaps for those long journeys at sea!—but a sealed plastic bag in the refrigerator is the way to go if you'd like to keep your stash fresh for anywhere from 1 to 2 weeks or longer.

Culinary Possibilities

India is the homeland of pickled limes, or lime pickle, whichever you prefer to call it. Spiced, brined limes were fashionable among British homemakers (Eliza Acton shares a recipe in 1845's *Modern Cookery for Private Families*). These bracing and salty concoctions are completely different from a finishing squirt of fresh lime juice, and rope in more of the flavor of the essential oils in the rind.

If you encounter many ripe limes at once, I recommend having a cocktail party. Crafters of fine mixed drinks love Key lime juice for its multidimensionality, but if you give a guest a drink with Persian lime juice they will knock it right back nonetheless, don't you fret. If getting your social circle blotto is not in the cards, make a batch of old sour. It's nothing but lime juice fermented with a few blazing-hot fresh chili peppers, and it slaps bland foods awake with just a few douses.

INDIAN LIME PICKLE

Makes about 3 cups (720 ml)

This is a burly condiment, not something you'd eat straight from a spoon. The first time I ever tried lime pickle I was at a fancy Indian restaurant, and I unwittingly spooned a giant blob of it onto my plate, thinking it was some kind of East Asian analog to salsa. It's not—it's bracing, bitter, and salty as hell. I spat it into my napkin then, but I love it now. The idea is to stir dabs of it into relatively rich or bland foods (a spoonful of lime pickle makes steamed basmati rice and a basic dal sing). I like mine just a little saucy, so I add tomato paste, but you can omit it for lime pickle that'll really put hair on your chest.

1 pound (455 g) limes (Key or Persian)
1½ tablespoons vegetable oil, divided
2 tablespoons kosher salt
1 tablespoon brown or yellow
    mustard seeds
1 (2-inch/5 cm) piece fresh gingerroot,
    peeled and chopped
2 teaspoons cayenne pepper
1 teaspoon asafetida, optional
¼ cup (55 g) brown sugar
1 (6-ounce/170 g) can tomato paste
2 tablespoons apple cider vinegar
2 tablespoons water

Remove the pips (the nodules at the stem ends) from the limes. Heat ¾ tablespoon of vegetable oil in a medium skillet over medium heat. Add the whole limes, and cook, shaking the pan from time to time, until the limes are brown in spots, about 5 minutes. Set aside to cool, then carefully cut the limes into 8 pieces per Key lime, or 16 pieces per Persian lime (be careful—the greasy limes can be slippery). Put the salt and lime pieces in a quart jar, screw on the lid, and shake to combine.

Screw the lid on fingertip tight, and let the jar sit in a sunny spot to cure for 1 to 4 weeks, shaking the jar every few days to redistribute the salt. Eventually a slightly gooey brine will form, and possibly not all of the salt will dissolve. Don't sweat it. Check on the jar after 1 week; it should develop a powerful but not off-putting smell somewhere between lemon furniture polish and lime candies. If it still smells fairly fresh, let it cure longer.

When you decide the limes are ready, finish the pickle. Put a medium nonreactive saucepan over medium heat. Add the remaining oil and mustard seeds, and cook, stirring constantly, until the seeds sizzle in the oil, about 2 minutes. Add the ginger, and cook, stirring, until aromatic, 1 to 2 minutes more. Add the remaining spices plus the sugar (it helps to have the fan on), and cook until the sugar is dissolved and the whole mess bubbles, about 1 minute. Add the limes and their juice plus any residue from the jar,

tomato paste, vinegar, and water. Cook, stirring, until the flesh of the limes breaks down somewhat, about 3 minutes.

Pack into a jar or jars. Insert a table knife along the sides of the jars to release big air pockets. Screw on the lids, then let the jars sit out for a day before refrigerating for up to months and months. The pickle is best after its flavor has developed for at least 1 week.

NOTE: Asafetida is a dry powder made from the gum of a rhizome that's mainly cultivated in India. It's kinda stinky, but its musk lends bland ingredients like potatoes or dal a little more dimension.

Old Sour

Makes about 1½ cups (360 ml)

Make a brine from the juice of the sourest of citrus, combine it with some of the spiciest hot peppers, and ferment it? Sign me up! This condiment is a classic of the Florida Keys, and it does not mess around. A little goes a long way, and it's an ideal match for the seafood dishes common in Florida and the Caribbean. I also like it on anything I squirt fresh lime juice over, but with salt and heat, too.

1½ cups (360 ml) fresh Key lime juice (from about 1½ pounds/680 g Key limes)

1½ teaspoons kosher salt

6 whole small hot red peppers or 2–3 habanero peppers, halved lengthwise

In a glass measuring cup, combine the lime juice and salt. Stir to dissolve.

Poke the peppers a few times with the tip of a paring knife, and slip them into one or two sterilized glass bottles. (I like to use old hot sauce bottles.)

Pour the lime juice into the bottles. Seal and ferment at room temperature for 3 to 4 weeks.

This is where it gets fun. Sprinkle a few dashes of Old Sour on some food, and give it a spin. If you'd like it to be a little spicier, drop another hot pepper in the bottle and let it ferment a few weeks longer. Old Sour keeps for ages, though its flavor will start to fade somewhat after 6 months.

ALSO TRY WITH: Persian limes, though the funky edge of Key limes is preferable.

LOQUATS

Eriobotrya japonica
Rosaceae family
Southern US in zones 8–11

A LOQUAT BY ANY OTHER NAME would taste as sweet, so take your pick: Japanese plum, Japanese medlar, Chinese plum, and japonica are some. They're known as *biwa* in Japan, where they are particularly beloved. Around New Orleans, old-timers call them misbeliefs, perhaps because of Italian immigrants who planted loquat trees there (*nespoli* is the Italian name; *nispero* is the Spanish).

Loquats originated in southeast China and today are grown in subtropical to mild climates the world over, from Israel to Brazil to Hawaii. Florida and California are big states on our small domestic loquat scene, but trees in Texas and warmer areas of the South can have loquat luck, too. The evergreens are popular as ornamentals because of their glossy green foliage and fragrant white blossoms. Some varieties are self-pollinating. Loquats do better some years than others, but the trees tend to produce quite a bit in the years when they fruit. In somewhat cool or hot and moist climates, a loquat tree will thrive but not bear fruit.

Their flowers were at one point used to make perfume, and one can enjoy the fragrance just being near a tree in bloom. The trees can grow up to 30 feet (9 m) tall, though that makes getting to the fruit on higher branches tricky. Some owners prune them to maintain a height of 10 feet (3 m) or so.

Loquats ripen on the tree, and the fruits that bats and birds don't get to, you will. If you enjoy a backyard nature show, a loquat tree is a big plus, because it's a magnet for fauna. Wildlife spread the seeds around. Seeds that wind up in your compost could very well become seedlings, so keep that in mind.

Loquats are about 2 inches (5 cm) long, on average, and range from yellow to golden orange in color. Their flesh is succulent and sweet, with a flavor like a cherry crossed with an apricot. Their skins are not objectionable, but I prefer to peel them.

At one point, Orange County, California, was home to one of the largest loquat groves in the world. Horticulturalist Charles Parker Taft planted them in 1888 and sparked a Southern California loquat fad that petered out in 1924; most of those trees were torn out to make room for more profitable oranges. Loquats are now a niche fruit, to the benefit of those who live near a person fatigued with managing their own loquat tree's bountiful output.

Loquats are a winter fruit, generally. Some varieties aren't ripe until late spring. By the way, loquats are not botanically related to kumquats, though both are small and orange-ish.

Harvesting and Storage

Loquats don't travel well. Refrigerating them will prolong their freshness, but in general don't dally in using them, as they bruise easily and will be fouled up with brown spots after a few days.

Loquats have one or more large seeds per fruit, but they release easily from the flesh. The seeds are mildly toxic. You'd have to ingest

an absurdly large volume to be poisoned, though it's not so much about eating the seeds as it is about using them to flavor other foods, as some do with stone fruit pits. In Japan, some people infuse the pits in liquor, and some preserve recipes call for tossing a few pits in while the fruit cooks.

For eating fresh, loquats are much better peeled. Cut one in half and you can slip it out of its skin as if it were a little jacket.

Culinary Possibilities

Loquats are best fresh. Dice them over cereal, cottage cheese, or a bowl of rice pudding. They kick butt in chilled and colorful fruit salads.

If you have access to a loquat tree, you likely have far more loquats than you could possibly eat fresh. For you bakers out there, try substituting loquats in recipes for apricots. You'll want to peel them, which will take forever but hopefully be fun and relaxing.

Loquats are high in pectin, so if you like to make jam, go for it! Loquat jam cooks to a deep, marmalade-y orange color and is quite handsome. Leave the skins on. For 3 pounds of seeded and quartered loquats, use 1 pound of sugar. Some cooks like to add a tablespoon or two of lemon juice.

Loquats dehydrate well, and their flavor concentrates into something between a date and a dried apricot. Simply quarter them, remove the seeds and as much of their white membranes as possible, and pop them in the dehydrator until pliable and leathery. As with jam, leave the skins on. Another way to burn through piles and piles of loquats is to make fruit leather, cutting it with some plain applesauce if you like.

HONORABLE
MENTION
★ ★ ★ ★ ☆

BEAUTYBERRIES

Callicarpa americana
Lamiaceae family
Southeastern US

The best thing about beautyberries is how unreal they look. These things are stunners, clustering around their stems like beads, their color ranging from magenta to purple. Compared with other purple berries, the cartoonish shade of beauty-berries is very un-berry-like. They could very well inspire a new Crayola crayon—both in name and color. As for their flavor . . . well, let's say their immediate charms may be predominantly visual.

American beautyberries are decidu-ous shrubs native to the southeastern states and can be found growing in shady forests from eastern Texas to coastal Maryland. They like hot, humid summers and mild winters. The berries start off as small purplish or white flowers that

grow into eye-catching drupes, becoming especially prominent around August or September. The berries will stick around on the stems for a while after the leaves have dropped. The ones I've sampled were somewhat astringent and neither repulsive nor compelling, but it was late in the summer and perhaps they weren't fully ripe yet. If you have beautyberries growing near you, keep sampling them as the season progresses; they may become juicier and sweeter. Sample small amounts at first; some people feel nauseated after eating them.

Native Americans used beautyberry roots, leaves, and berries for various medicinal purposes, mostly as boiled infusions and teas. Crushed leaves have compounds that repel insects, and settlers would rub them on their skin to soothe itches and keep mosquitoes at bay.

Among those who appreciate beautyberries for eating, jam or jelly is the go-to preparation. You'll need to add pectin. I've sadly not had a chance to make it myself, but if you want to give it a shot, wild plant expert Mark "Merriwether" Vorderbruggen, author of the *Foraging Texas* blog, compares its flavor to "rose petals and champagne." What an enticingly luxe description! His formula calls for you to boil 1.5 quarts of berries and 1 quart of water for 20 minutes, then strain out the solids. Add 4.5 cups of sugar and 1 envelope of powdered pectin to the strained liquid. Return the liquid to a boil and cook for 2 minutes, skimming off any foam before pouring into sterilized jars and canning.

You may find white beautyberries, a variant of *C. americana*. These are also edible, although for some reason the colorful flair of the purplish berries seems to make them more worthwhile to chase down if you're going to be making preserves.

Alas, American beautyberries don't grow where I live, but smaller Asian beautyberries (*C. dichotoma*) do. They are nontoxic but unpalatable. You can tell the difference not just from the size of the drupes, but by how the berries grow: *C. americana* grows directly from the main stem, while *C. dichotoma* dangles in petite clumps from multiple tiny, thin stems. Both American and Asian beautyberries are popular and highly attractive landscaping plants. If you're fortunate enough to encounter American beautyberries and their flavor does not lure you, leave them for the birds to feast upon and simply appreciate these handsome plants for their good looks.

Loquat Ambrosia

Serves 4–6

This contemporary version of the coconut-laced fruit salad you might recall from church potlucks or big community buffets is a smart way to introduce loquats to skeptics. You won't find any miniature marshmallows here, but no one's forcing you not to add some if you feel compelled to.

1 orange
1 cup green grapes, halved
1 cup pineapple chunks
1 cup peeled loquat chunks
Finely grated zest of ½ lime
½ teaspoon honey
¼ cup (20 g) unsweetened coconut
    flakes, plus more for garnish
¼ cup (60 ml) heavy cream
1 banana, peeled and sliced, optional
¼ cup (30 g) toasted pecan pieces

Cut the top and bottom off the orange. Sit the orange up on its cut bottom and, starting at the top, work around it to cut off strips of the remaining peel. Discard the peels.

Hold the orange over a medium bowl, and carefully slide a paring knife between the flesh and the membranes to free the orange segments from the membranes, letting the segments fall into the bowl. Squeeze any remaining juice out of the orange membranes, and discard the membranes.

Add the grapes, pineapple, loquats, lime zest, honey, and coconut to the bowl, and toss to combine. Cover, and refrigerate until cool.

Just before serving, whisk the cream in a large bowl until soft peaks form. Fold the cream and the banana (if using) into the bowl of cold fruit. Sprinkle the pecans and a little more coconut over the top, and serve immediately.

NOTE: To toast pecans, bake them on a rimmed baking sheet in a preheated 350°F (175°C) oven for 3 to 5 minutes, until aromatic. Cool before adding to the salad.

ALSO TRY WITH: Pitted, chopped unpeeled apricots in place of the loquats.

MAHONIA / OREGON GRAPE

Mahonia aquifolium
Berberidaceae family
Western US and Canada

A MEMBER OF THE BARBERRY FAMILY, this shrub is called mahonia, holly-leaved barberry, Rocky Mountain grape, and plenty of other names. I prefer Oregon grape because I used to live in Oregon, and it pulls at my sentimentality. Oregon grape is the state flower of Oregon, in fact. They are dainty yellow flowers, and a cheerful sight in the spring.

With glossy, hollylike leaves, Oregon grape is easy to spot even when it's not blooming or fruiting. Resilient and attractive, it is a popular ornamental in yards and public landscaping. It grows between 3 and 6 feet (0.9 to 1.8 m) tall, and makes hearty hedges and borders.

All of this is good news for restless berry pickers of the urban wilds. If you need a fruit foraging fix and you're near the Northwest, Oregon grape can't be too far away.

In the late summer and early fall, the small berries ripen to a purple-blue and have a powdery, grapelike bloom. They grow in clusters, but those pesky barbed leaves present an obstacle. Some report they taste good raw, but I disagree. If they tasted better, way more people would be out picking them. Seedy, tart, and bitter, they are a face-scrunching berry for sure. Tribes in the Northwest used multiple parts of the plant for medicinal purposes, and herbalists continue to employ it in various forms. I, however, value Oregon grapes as something you can smother with refined sugar to make a damn fine jam.

In classic fashion, I didn't know about Oregon grapes until moving back East, so I sought them while visiting family near Portland one summer. I did not need to look far—in fact, all it took was crossing the street. I picked my way along the parking lot by the condos to the nearby city park and continued working my way down; Oregon grapes grew like a plant pathway. I grabbed some salal berries for good measure, too, and in an hour I had enough to make a batch of jam. It was

grapey indeed, and nearly as dark as night. I would have happily spent the rest of the trip in a flurry of picking and preserving, but our agenda had other ideas, and I had to be satisfied with the two jars of jam I did make.

M. swaseyi, or Texas barberry, is a similar shrub, but its fruit is an edible red berry. To make things especially confusing, it is sometimes called Texas Oregon grape. You can use the berries in jam or jelly just as Oregon grapes, but it might not be as pretty. Neither grows where I live now, so I'll have to plot my next visit to the Pacific Northwest accordingly, allocating for time to walk across the street and hustle up enough berries for another couple of jars. Proximity is everything with foraging, even if you fly across the country to call yourself near.

Harvesting and Storage

These berries are fairly sturdy and able to withstand some knocking around in a bucket. They are also small, so getting a quantity takes a chunk of time. Once the berries get that nice bluish color, it's tricky to tell when they are ripe. You can start picking in midsummer, but waiting until the berries are slightly softer makes for a better flavor. Also, they don't all ripen at the same rate on the same plants. But I picked some that were pretty firm, and they still made good jam.

To help remove any small stems, shake the berries around in their bucket or box, which will knock them loose. Or you could try the elderberry trick of covering the berries in water and picking out the debris that rises to the surface.

Culinary Possibilities

Jelly or jam is all I've heard about, outside of homeopathic applications.

Oregon Grape Jam

Makes 1–2 half-pint jars (240–480 ml)

This recipe is scaled with my patience for picking Oregon grape berries. I made a batch the last time I visited Oregon and gave it as a wedding present to my sister and brother-in-law, along with some pot holders I'd sewn. It was a pretty chintzy wedding gift, but even so, it was hard to part with. I should make them more jam.

2–4 cups Oregon grape berries
2–3 cups (400–600 g) sugar

In a medium saucepan, add enough water to barely cover the berries. Bring to a boil and cook, stirring occasionally, until the berries are fall-apart tender, about 10 to 15 minutes. Run the mixture through a food mill set over a large bowl to remove the seeds.

Discard the seeds, and measure the strained pulp. Add 1 scant cup (about 190 g) sugar for every cup of pulp. Return to the saucepan and cook, stirring constantly, until very thick. Divide between sterilized half-pint jars. If you like, seal and process in a water bath canner for 10 minutes. Otherwise, cool and refrigerate your jam for up to 2 weeks.

NOTE: In a small batch, it's easier to cook the jam down until it's quite thick. Add pectin if you like, but I don't think it's necessary. Your yield on the pulp may vary depending on the mix of berries you use.

SALAL

Gaultheria shallon
Ericaceae family
Western US and Canada

Salal is a low evergreen shrub that produces dark blue berries that were, and are, important to Native Americans of the Pacific Northwest. It's an attractive plant, with bright green leaves and small white flowers that droop from pinkish red stems, but the berries have a bit of a hairy look that reminds me of testicles. Maybe that's why more people don't pick them nowadays.

Like Oregon grapes, salal berries are finger stainers of the first order. Their blossom ends have a distinct five-pointed star pattern that keeps them from looking *too* much like testicles. They're about the size of blueberries, and sweeter than Oregon grapes. Opinions on their flavor range from sweet to mild and flat. It's likely that differences in the character of one plant

to another account for this as much as personal preference. Some people love to eat them off the plant, while others make the berries into jams or fruit leathers.

There's a small art to picking salal berries, which do not like to give up their stems. Go for the darkest ones, which will be the most flavorful. To keep them from squishing, lightly grip one between your thumb and forefinger and roll or twist it from the stem. You can also snip off the entire long cluster with scissors and deal with stemming them at home.

Salal berries are of no matter to consumers of its primary commercial use: greenery for floral arrangements. The shapeliness of its leaves and their ability to hold up well in a vase make it a favorite of the floral industry.

Foraged Jams

There are very few foraged fruits you can't make into jam or jelly. Whether it's worth the effort to do so is up to you. Such is the crapshoot of foraged fruit preserves.

Some fruits taste bland raw, and some taste just plain nasty. When cooked down with sugar, though, an amazing transformation can happen. Colors deepen and flavors pop, plus you get that gorgeous thick-but-spreadable consistency and shimmery look. The whole process enchants me, even though I may curse at how hot it makes the kitchen or how many pounds of sugar I am ripping through or how canning reduces my efforts to this modest little row of pretty jars lined up on the counter after what feels like hours and hours of work.

If you have a relationship to canning fruit preserves, perhaps it is not as masochistic as mine. The important thing to keep in mind is that even experienced jam makers run into speed bumps from time to time. These can only make a successful batch seem all the more triumphant.

Anymore, I like to make fruit preserves in small batches—two to five jars or so. A smaller volume of fruit reaches the gelling point faster, and sometimes a foraging haul might be too small for a big batch. Besides, an unsuccessful small batch of fruit preserves feels like an interesting experiment, while a big batch gone wrong can wreck a whole day.

MAYHAWS

Crataegus spp.
Rosaceae family
Southeastern US

MAYHAWS ARE TREES OF RIVER, swamp, and creek. Native to acid alluvial soils from North Carolina to Florida and as far west as Texas, they are a cherished emblem of traditional southern foodways. So many fruit trees have names that are goofily misleading, but mayhaw is spot on: The fruit ripens in May, and they are a species of hawthorn.

Culinarily speaking, mayhaws are the shining stars of hawthorns. Humans have been using the fruit of hawthorns as a food source for millennia. Though generally not tasty, most hawthorns bear abundant but bland and mealy little fruits, which were relied on in lean times, specifically in North America. Like apples, hawthorns have seeds that, if eaten in isolated quantity, could cause illness or death. But if the seeds were so dangerous, likely you and I wouldn't be here today; hawthorn seeds have been found in archaeological sites of early human dwellings on various continents. Just as we eschew apple seeds, our ancestors probably devised workarounds, too.

Mayhaws are comparatively larger than most North American hawthorn fruits and pack a punchy tartness. They are never eaten out of hand, or if they are, then that person has no taste buds. Mayhaws are raw material for juice, pulp, and especially jelly, all of which have a crab apple / rose hip flavor. The fruits are the size of large crab apples and vary in color from yellow to ruby.

To gather the fruit in the wild, fallen mayhaws floating in standing water were collected—it's a southern version of the flooded cranberry bog method. Foragers would shake the trees to loosen the fruits. "I have Mississippi childhood memories from 50 years ago of Grandmother and me going down the road to the low land where the rainwater would stand in murky blankets around the feet of the mayhaws," wrote Sherry Pendleton in an issue of *Countryside & Small Stock Journal*. "The tiny apple-like fruit would hang in ruby spheres from the small understory trees. Papa's rubber boots slopped up and down on my bare feet as I waded to scoop up the fruit with a colander as the little apples floated on the water." People would even use boats to get to mayhaw stands along creeks and rivers, using nets to retrieve the fallen bounty from the surface of the water.

The modern story of the mayhaw is one of cultivation making accessible what was once wild. People have grown mayhaws for pleasure since the late 1800s, but because the natural habitat of the mayhaw is shrinking, fruit growers have stepped in to keep it going strong as a traditional southern food. Even so, diminishing stands of wild mayhaws means a loss of something besides what can be eaten.

The western mayhaw (*C. opaca*) grows from eastern Texas to Alabama, and the eastern mayhaw (*C. aestivalis*) from eastern Alabama to central Florida and Virginia. Their characteristics as a food crop and ornamental are similar. The trend of mayhaw enthusiasts and farmers growing trees as ornamentals and small-scale orchards makes the

fruit more available, in terms of both harvesting and quantity. In orchards tarps stretched out under the branches stand in for the water to catch the ripe fruit.

With the shift from wild to farmed habitat has come a surge in regional mayhaw pride. Gift shops sell jars of mayhaw jelly to tourists and aficionados, and multiple mayhaw festivals are now held in Georgia, Texas, Louisiana, Arkansas, and Florida. Mayhaw jelly looms large at these, as it is the most crowd-pleasing and storied mayhaw preparation.

It is becoming more likely to encounter a cultivated mayhaw tree than a wild one. I wonder if our hunger to continue developing swampy wild areas of the South will make cultivated mayhaws the equivalent of polar bears in zoos. It's better to have those mayhaws than none at all, but it's best to have solid futures for both.

Harvesting and Storage

Mature mayhaws are the ones you want. Their color will be bright, and their flesh will be somewhat soft. As hawthorns, mayhaws have thorns like burly needles sticking out from their branches, but if you are using the old tarp trick to harvest them, the thorns will not get you much.

A gatherer of mayhaws has many options for processing them. Fresh mayhaws keep decently under refrigeration for a few days, or frozen for a few months. The storage method most convenient for later use is to simmer the mayhaws in a little water and extract the juice, which can be frozen or made into jelly.

While it is possible to freeze mayhaws whole, they are almost always cooked so the juice can be strained out. The juice can be frozen to add to recipes (such as cocktails, punch, syrup, or vinaigrette) or transformed into jelly. If you are inclined to experiment and not the sort of person for jellies or punches, then go rogue and think of other ways to showcase this beloved fruit, the Queen of Haws.

Culinary Possibilities

The skin of mayhaws is bitter, so recipes devise ways to employ strained, cooked pulp or juice. Mayhaw pulp has its fans and uses, but mayhaw jelly and juice are the handiest and most popular starting points for mayhaw recipes. Whether for desserts—cheesecakes and pound cakes laced with mayhaw juice—or salad dressings and glazes for meat, mayhaws work best in food where the energizing interplay of bright and sugary tastes makes sense.

Mayhaw Juice

Makes 12 cups (3 L)

This is step one for making mayhaw jelly, but mayhaw juice has other uses. Think of this tart liquid as a base for syrup, beverages, and sauces. The method below comes from the Louisiana State University Agricultural Center.

16 cups (4½ pounds/2 kg) mayhaws
12 cups (3 L) water

Rinse and sort through the mayhaws, discarding any debris but leaving on the stems and blossom ends. Combine the mayhaws and the water in a large, heavy-bottomed pot. Simmer for 30 minutes, cool, and strain twice: once through a colander, and a second time through a jelly bag or a colander lined with a few layers of cheesecloth. Don't throw out the sediment that settles to the bottom of the strained liquid.

Freeze the juice in quart containers for easy portioning, or can in hot, sterilized pint or quart jars, leaving ¼ inch (6 mm) headspace. Seal and process in a water bath canner for 10 minutes.

NOTE: If you need to scale this down, use this formula: For every 2 quarts (2 L) of fruit, add 6 cups (1.5 L) of water.

MEDLARS

Mespilus germanica
Rosaceae family
Zones 5–8

Dull brown and only delectable when halfway rotten, medlars resemble a few things prudish people do not put their mouths anywhere near.

I have never set eyes upon a real medlar tree, but I love the idea of medlars. They are divinely ugly with their massive calyx (blossom end) and spotty brown skin. They are the size of small apples or persimmons, and like persimmons they need to be bletted, which is a fancy fruit word for allowing something to ripen enough that it partially rots. This can be done either on the branch or off the tree, in the case of medlars. Lay them in a single layer in a cool place for a few weeks, and when they are wrinkled and squishy enough to be easily ruptured with light pressure, they're there. Squeeze out the mushy flesh, and leave the skins behind. Descriptions of the medlar eating experience often evoke cinnamon and wine-tinged applesauce, but grittier. Think of it as fruit butter that you don't have to make yourself. Well, then, how to use these things? Recipes call for adding medlar pulp to custards for tarts, or boiling the fruits whole to make jelly and jam.

The medlar is a bizarre relic of a fruit. It originated in Southwest Asia and southwestern Europe. There was a saying in France, *à la Saint-Simon, le fruit du meslier est bon*, referring to St. Simon's October 28 feast day, a time when medlars were likely ready to eat. Shakespeare used medlars multiple times in his plays to comically allude to taboo parts of the human anatomy.

The trees are more conventionally attractive than the fruits. They don't grow very tall and are self-fertile, so they are well suited for planting as ornamentals. You are more likely to find a medlar tree in North America than a unicorn, but medlar trees are uncommon enough that the quest for either can feel fantastical. If you see one, it is probably on the property of a fruit nerd. You should know that person, so go introduce yourself. They are probably dying for someone to geek out on medlars with.

Mayhaw Candied Sweet Potatoes with Chipotle and Bacon

Serves 6 as a side

The dressing for this warm potato salad is sweet from a big blob of Mayhaw Jelly, but bacon and smoky chipotle pepper cut through it. Serve this at Thanksgiving to liven things up, or any other time of the year.

4 slices bacon
2½ pounds (1.1 kg) sweet potatoes, peeled and cut into 1-inch (2.5 cm) cubes
Salt and pepper to taste
3 cloves garlic, minced
¼ cup (60 ml) Mayhaw Jelly (page 220)
½ canned chipotle pepper, minced
1 tablespoon apple cider vinegar
⅔ cup (80 g) pecans, toasted and chopped

Preheat the oven to 425°F (220°C).

Put the bacon in a heavy skillet over medium-low heat and cook, flipping occasionally, until crispy. Set the bacon on paper towels to drain, then crumble into pieces; set aside. Pour the bacon grease in a small bowl, and keep the skillet handy.

Put the cubed sweet potatoes on a rimmed baking sheet. Toss with 2 tablespoons of the bacon grease. Season with salt and pepper, and roast, shaking the pan after 15 minutes. Continue roasting until the potatoes are slightly browned at the edges and cooked all the way through but not so soft they fall apart, 15 to 25 minutes longer.

Meanwhile, return the skillet to medium heat. Add 1 tablespoon of the bacon grease; when hot, add the garlic, and cook for 1 minute. Add the jelly and chipotle, and cook, stirring, until the jelly is melted. Add the vinegar. Season to taste with salt. Keep the dressing warm over the lowest heat possible.

Turn the cooked sweet potatoes into a large bowl. Toss with the warm dressing, all of the bacon crumbles, and half of the pecans. Top with the remaining pecans. Serve hot or warm.

NOTE: For a vegan version, omit the bacon and use olive oil instead of the bacon grease. If you'd like a smokier taste, use smoked almonds instead of the pecans.

ALSO TRY WITH: Habanero Crab Apple Jelly (page 102) or any tart jelly.

Oh Nuts!

Every forager draws the line somewhere, and for me it's nuts. I don't forage for nuts, because I love fruit and channel my available patience to using it. I don't have patience left for nuts.

As you go out on your fruit adventures, you'll doubtlessly run across many forageable tree nuts: hickory, beech, walnuts, acorns, chestnuts—the list goes on. If you are so compelled, you can widen your net and start identifying and gathering those, too.

I was telling a friend about a nonforaging hobby I'd taken on: bokashi fermentation, an anaerobic way of pickling food scraps with a specific blend of microbes so they'll easily break down in soil. With this method, you can compost meat scraps, cooked grains, and even oils. It fascinates me, and I'm always trying to bring on new converts.

"Why would I want to bother with that?" replied my friend, who has a composting toilet. It struck me that poop is plenty enough for one household to compost. Likewise, just because you *can* forage for something does not mean you *have* to. There's only so much of you to go around. It's fine to be choosy! I choose fruit, leaving nuts for others to have fun with.

Mayhaw Jelly

Makes 6 half-pint jars (1.5 L)

"The mayhaw is most often used in jelly, which we eat for pleasure rather than its nutritional value," says a handout compiled by the Louisiana State University Agricultural Center, which shared this recipe. Jiggly and ruby-colored mayhaw jelly is a state jelly of Louisiana, both symbolically and officially (the state legislature decreed it this in 2003; it shares this distinction with sugarcane jelly). It tastes like a tart honeyed apple and rocks all the typical jelly applications, but it's versatile in recipes as well. Dole the precious substance out in dibs and dabs.

4 cups (960 ml) strained
  Mayhaw Juice (page 216)
1 (1.75-ounce) box (5 tablespoons)
  powdered pectin
5½ cups (1.1 kg) sugar

In a heavy-bottomed Dutch oven or small stockpot, mix the Mayhaw Juice and pectin together. Quickly bring to a hard, rolling boil, stirring occasionally. Add the sugar, and stir until the pot returns to a full boil. Boil for 1 minute and 15 seconds, stirring constantly. Remove from the heat, and skim off any foam. Pour into sterilized half-pint jars, leaving ¼ inch (6 mm) headspace. Seal, and process in a water bath canner for 10 minutes.

NOTE: Almost all mayhaw jelly recipes call for pectin, though mayhaws have enough pectin to set without it. You can omit the pectin, but the jelly will need to cook longer to set.

MULBERRIES

Morus spp.
Moraceae family
Throughout the US and Canada

MULBERRIES ARE TREES OF MEMORY, if you happened to grow up around one. There will never be mulberries sweeter and juicier than the ones gathered from a tree that grew where you played during your childhood.

That's the rub with mulberries, for in adulthood they never fully live up to our expectations. Other berries are more flavorful and juicier, particularly blackberries and raspberries—which mulberries resemble at first glance. Mulberries are not related to caneberries, though, and forever mulberries will be poseurs, early-summer stand-ins for the blackberries and raspberries yet to come.

I still love mulberries. They are fun to gather, and one must admire their productivity. In a good year a single mulberry tree will provide enough easily harvestable fruit for a procession of pies. And that's even after the birds have feasted. Mulberries give and give, to the consternation of homeowners who prefer their driveways unstained and sidewalks free of sticky smears. Don't park your car under a mulberry tree when it's fruiting. The berries will stain your car, and the birds in the branches who've been gorging on the berries will also unleash their berry shits on it. (When mulberry-tainted bird poop lands on the sidewalk, it's a swirly lilac-white, pretty in its way.)

You will find mulberries of different colors. The red mulberry (*M. rubra*) is the only one native to North America, and when ripe its fruits

are a very dark purple-black. Its fruits are virtually indistinct from the color of ripe black mulberries (*M. nigra*), though black mulberries tend to bear the largest fruits. The white mulberry tree (*M. alba*) is much more common in North America, though. Don't put too much stock in its name, as it can bear fruit that's either purple-black or white when ripe. So confusing! The color of the berries will depend on the individual *M. alba* in question, as will their flavor. The tastiest mulberries I've sampled were the white-when-ripe kind, as were the most underwhelming. Don't give up on mulberries of any color if you sample a dud. Keep on looking, and keep on tasting.

M. *Alba* was introduced in a short-lived craze for cultivating silkworms, an interesting footnote in American history. The hope was to jump-start a profitable silk industry, beginning by planting white mulberry trees to host silkworm larvae. Schemers hoping to get rich planted mulberries in orchards and hedges. Once the trees were mature enough to host the larvae, the labor required to spin the silk outweighed the quality of the final cloth itself, and our fledgling silk industry fizzled out, beginning with the financial Panic of 1837. You'll now find *M. alba* (most of them with dark-when-ripe berries) coast to coast, and white mulberries are often classified as invasive.

While mulberries are not a commonly recognized fruit, they are familiar to some as place names. There's a Mulberry Street in my town, a quiet alley with brick pavers. It only spans a block, and I've walked down it dozens of times looking for the eponymous mulberry trees. I've spotted two crab apple trees and a grape arbor, but alas! No mulberries. I imagine they were cut down years ago.

The dark mulberries you are most likely to come across will stain hands and fingers a purplish blue-red; once I was out gathering mulberries with my dog, and the fringes of his white fur were purple for a few days (he'd sat in a pile of fallen berries). Not surprisingly, mulberries make a fantastic natural pigment for dyes and inks. The legend of the lovers Pyramus and Thisbe contains an origin story for the black mulberry tree along these lines: It started out as a white mulberry tree in a cemetery where the couple had planned to meet. Thisbe arrived first, and startled at the sight of a lioness, she fled, snagging her veil and scraping herself on the mulberry tree in the process. When Pyramus arrived shortly after, he saw the veil, presumed the lioness had gobbled up Thisbe, and killed himself. When Thisbe returned, she offed herself, too. Such a dramatic bloodbath stained the mulberry tree forever.

My favorite mulberry trees grow in a cemetery, as it turns out. I am there to meet no forbidden love, and the most threatening animals there are not roving lionesses, but groundhogs who have dug vast networks of burrows. I'm afraid my dog may get in a scrap with one, so I keep him close on the leash as I collect berries under the gentle shade of the trees.

Harvesting and Storage

It's very possible to just walk up to a mulberry tree and pick the ripe berries. What's faster—and more fun, in my opinion—is to get down on the ground and gather up the ripe ones that have already dropped (wear shoes whose soles you don't mind getting gooey and purpled). Inevitably more ripe berries will plummet down as you do this, making tiny thumps to punctuate the backwash of bird sounds and cars driving by.

That method is only fun if the tree grows on a softly tufted spot. My favorite mulberry trees are carpeted below with moss and creeping Charlie, and rifling through their greenery is soothing.

For efficiency, bring a sheet or tarp along and spread it on the ground beneath the branches. Shake the branches, and the ripe berries will rain down. Don't pile them up too high in the pail or bin, because the pressure of the top ones will mash down the bottom ones.

Use mulberries quickly; they bruise easily, which is one reason you don't see them in grocery stores. Also, if you've plucked them from the ground, sort through them carefully; I've pulled a few sneaky mulberry-sized rocks out of my berry clutches before. You may notice some tiny critters of the baby insect variety venturing out from the mulberries' crevices. Don't think about it too much, and just chalk it up to extra protein.

Mulberries freeze well. Stockpile a whole bunch if you love baking pies, and draw upon them in the winter, to remind you of sun-dappled afternoons gathering them in your sandals.

Mulberries have a tiny segment of a green stem sticking from their tops. These things look like minute inchworms and are cute in their way. There's no need to trim them off, as doing so will madden you unnecessarily. I do find the stems visually distracting, though, so think of ways to disguise them: Bake mulberries in muffins, hide the stems under the top crust of a pie, or infuse mulberries in a kompot or shrub that will be strained later.

Culinary Possibilities

Besides their delicacy, the other reason you don't see mulberries in grocery stores is that they taste underwhelming, a bit wan and vegetal. They are enough like blackberries to remind us of blackberries, but they will never be blackberries (the Who didn't include a verse about mulberries in the song "Substitute," but they very well could have). Also, once cooked down, mulberries have a watery texture that makes them unsuitable for quickly cooked compotes and sauces. I've never made mulberry jam (the stems are texturally problematic), but if you do, you are best off adding pectin.

Good white mulberries are wonderful right from the tree, but I've never cooked with them (partly because my favorite white mulberry tree is ringed with poison ivy and hard to get close to safely). I'm not sure if their more honeyed, floral flavor would do well in pies and jams, or what white mulberries would look like after cooking—a pale, sludgy mess?

What you do with mulberries will depend on how you feel about them. A sentimental attachment is very helpful in enjoying them to their fullest, as is a heavy hand with sugar and spice (mulberries need all the help they can get). A squirt of citrus plays up their fruitiness.

You are best off thinking of mulberries as a team player with other, more flavor-forward fruits. Rhubarb's tartness is a fantastic partner for mulberries ("The berry-rhubarb team makes one of the most economical farm fruit pies—and one of the best," according to *Farm Journal's Complete Pie Cookbook*).

MULTIGRAIN MULTI-BERRY MUFFINS

Makes 12 muffins

As much as I love pancakes, it's so easy to stir together a bowl of batter and stick a bunch of muffins in the oven—it frees me up to read the newspaper instead of standing over a griddle.

1 cup (145 g) whole wheat flour

¾ cup (70 g) oat flour

¼ cup (30 g) ground flaxseeds

1 tablespoon baking powder

½ teaspoon salt

½ cup (100 g) granulated sugar

Finely grated zest of 1 orange

½ cup (120 ml) orange juice

1 cup (240 ml) unsweetened
    nondairy milk

⅓ cup (80 g) melted coconut oil

1½ teaspoons vanilla extract

2 cups (290 g) ripe but sturdy berries,
    such as mulberries or blackberries

Preheat the oven to 375°F (190°C). Grease a 12-cup standard muffin tin.

In a large bowl, whisk together the flours, flaxseeds, baking powder, salt, and sugar.

In a medium bowl, whisk together the orange zest and juice, milk, oil, and vanilla. Add the flour mixture, and fold with a rubber spatula just until combined (a few small lumps are okay). Fold in the berries.

Divide the batter among the prepared cups, filling each no more than two-thirds full (I use an ice cream scoop). Bake until the centers of the muffins spring back when lightly pressed and a toothpick inserted in the center comes out clean, about 20 minutes. Cool the tins on racks 2 minutes, then remove the muffins from the tins.

NOTE: You can line the tins with paper muffin cups, but I prefer the texture of muffins baked without them.

ALSO TRY WITH: Blueberries, huckleberries, blackberries. You could also throw in a few fresh cranberries, if you like, or substitute some raspberries for a portion of the other berries (an all-raspberry muffin will wind up too dense).

Mulberry–Strawberry Shrub

Makes about 1 pint (480 ml) shrub syrup

Shrubs originated as ways to utilize the abundance of seasonal fruit crops in the days before refrigeration or canning. Fruit and sugar were fermented together to make a slightly sour drink. Nowadays the common technique is to skip the fermentation and instead add vinegar. You don't want to use harsh, distilled white vinegar in a shrub. Softer, less acidic vinegars like champagne vinegar, rice wine vinegar, or most any fruit vinegar work well.

Mulberries impart a dark jammy color and cut the cloying brightness of other fruits (I prefer to mix berries when making shrubs). Strawberries are around when mulberries are ready to harvest, and they make a logical pairing here. Since the character of mulberries can vary so much from tree to tree, your shrub will always be on target; the glory of shrubs is that you can easily tinker with the sugar and vinegar levels to make a drink that's just to your taste.

2 cups (290 g) mulberries
2 cups (290 g) strawberries,
    raspberries, or blackberries
2–3 cups (400–600 g) sugar
¼–½ cup (60–120 ml) vinegar (I like
    rice vinegar, champagne vinegar,
    or white wine vinegar)

Rinse off the berries. If you're using strawberries, cut them into quarters. Toss the berries and sugar together in a large glass or ceramic bowl, and either refrigerate overnight or let the mixture sit on the counter at room temperature for up to 4 hours. You'll know it's ready when all of the sugar is dissolved and the fruit is slouchy and soft. If the sugar is not fully dissolved, just give it a stir and let it macerate for another few hours, or up to another full day.

Strain the liquid from the fruit through a fine-mesh sieve or a colander lined with cheesecloth set over a bowl. Press down on the fruit to release as much of the liquid as you can. It will be sticky and a little syrupy. Some fruits give off more liquid than others, so your yield here could vary quite a bit.

Add the vinegar, starting with the lesser amount. The shrub will be very intense, and possibly a little harsh, when you taste it right away. No worries—think of it as a concentrate, a base to be diluted with ice, chilled water, or even booze. I like to let my shrubs mellow in the fridge overnight so all of the flavors can settle in. Your final result should be puckery and jammy.

The following day, taste the syrup and add more vinegar or sugar, if necessary. To serve, pour over cracked ice and add a little water or soda water.

The undiluted syrup will keep, refrigerated in a tightly covered glass jar, for about 1 week. You can also freeze the syrup for up to 6 months.

ALSO TRY WITH: Blueberries, blackberries, raspberries, peaches, plums, nectarines, cherries, or grapes. Even loquats would be good in a shrub!

OSAGE ORANGES

Maclura pomifera
Moraceae family
Throughout the US

Eating an Osage orange won't make you sick, unless you're allergic to latex. Even if you're not, it's unlikely you'd eat one in the first place, as their flavor and texture fall somewhere between unpalatable and underwhelming. Despite their inedibility, I love the things. Bulbous and bright yellow-green, they look like props from a low-budget science fiction movie—like the brains of aliens, or the eggs of space monsters.

Osage oranges are not citrus. They are also not apples, despite their other nicknames: horse apples and hedge apples. The trees themselves are very common in the eastern United States, both in deciduous forests and as border trees on farmland (thus the *hedge* in hedge apples). The trees are not especially pretty but full of character, have nasty thorns on their branches, and grow quickly.

The fruits range between 2 and 5 inches (5 and 13 cm) in diameter. In the woods you'll come across them under the trees, as if someone dumped a giant bucket of misshapen tennis balls. Old-timers say that Osage oranges, if placed strategically on storage shelves, will keep spiders away. It turns out this is (a) not true and (b) perplexing, because spiders are awesome and eat pesky insects. Why would you want to keep them away?

The seeds of an Osage orange are edible and palatable. They somewhat resemble

sunflower seeds, but there are only fifteen or so per fruit, and they are embedded deep in its center, under inches of dense, fibrous flesh. Extracting them is thankless work for a meager, bland snack.

If I see Osage oranges on the ground while on a fall woods walk (scattered on the forest floor, they resemble tennis balls) and I have a bag with me, I always haul some home. Then I line them up in the railing of my front porch as a decoration—they'll last for a few weeks before brown spots start plaguing them. Inevitably the squirrels will come sneaking around and foul up my arrangement (if it's been a lean year for desirable nuts, squirrels will be desperate enough to chew into those buggers to get at the seeds).

The exterior of Osage oranges can be sticky; that's the latex, which is sometimes present to the point that you can actually see white drips of it. A lady selling Osage oranges as a novelty at the farmers market told me that she sometimes cut them into thin cross sections and dehydrated them. The disks shrivel up to form something vaguely resembling a flower blossom. I tried this and it worked, but I had to rub Goo Gone on my knife to get the cling of latex off. I also took a tentative nibble of a dehydrated slice, and even in that form, I was not compelled to eat any more. But I'll continue dragging them home, just because I delight in their otherworldly vibe.

OMG Maggots!

Inevitably a forager will have some uninvited guests tag along home with their harvest. A few stray bugs are no big deal, and teensy inchworms are kinda cute, but the sight of maggots can induce shudders.

Even if you sort through wild berries very well, sometimes a few little dudes will escape your scrutiny. I once unscrewed the lid on one of my batches of cherry bounce as it steeped in its jar and saw a dozen tiny, sliverlike maggots floating on the surface. They were dead, of course, pickled in fruity booze. Did I pitch it? Was it ruined? Nope, and you wouldn't need to toss out yours in a similar situation. A

modest scattering of maggots will not make you or anyone else sick. The FDA allows commercial canners to include one or more fruit fly maggots per 100 grams of tomato sauce, because fruit flies looove tomatoes. If you eat canned tomatoes, you have surely consumed maggots and lived through it. It's the old "great extra protein" joke. If you like, simply remove the maggots from your cherry bounce and press on.

Likewise, stray maggots that may be cooked into preserves won't ruin your batch. But if you are trimming fruit and see any infested with maggots, discard those pieces.

Mulberry and Peach Cobbler with Almond Topping

Serves 8

Compared with a deep-dish fruit pie, this bakes a little faster because the filling isn't as deep. It might look shallow in the large dish, but you wind up with the ideal 1:1 ratio of topping to fruit filling in every bite—no one component overpowers the other. The topping bakes up tender, thanks to the almond flour. The sliced almonds add a pleasing crunch—you can skip them, if you like, but it won't look as pretty or taste as exciting.

For the filling

⅔ cup (130 g) granulated sugar

2½ tablespoons cornstarch

Finely grated zest of ½ lemon

¼ teaspoon almond extract

3½ cups (505 g) mulberries
    or blackberries

2 cups (370 g) pitted, peeled,
    and sliced peaches

For the dough and topping

½ cup (65 g) unbleached
    all purpose flour

½ cup (65 g) almond flour

2 tablespoons cornstarch

1 teaspoon baking powder

¼ teaspoon salt

6 tablespoons cold unsalted butter,
    cut into small pieces

½ cup (120 ml) milk

1 egg white

¼ cup (60 g) turbinado sugar

⅓ cup (40 g) sliced almonds

Preheat the oven to 350°F (175°C) and position a rack in the center. Grease a 13 × 9-inch (33 × 23 cm) baking dish.

To make the filling: In a large bowl, stir together the sugar, cornstarch, and lemon zest. Fold in the almond extract, berries, and peaches. Scrape into the greased dish. Set aside.

To make the dough and topping: In another large bowl, stir together the flours, cornstarch, baking powder, and salt. Add the butter, and work with your fingertips or a pastry cutter until the mixture looks like coarse bread crumbs. Add enough of the milk to make a soft dough that's just a little looser than biscuit dough. Drop by dollops over the fruit in the dish (it'll look like there's not enough dough, but it'll puff up, don't worry).

In a small bowl, beat the egg white with a fork just until frothy. Stir in the turbinado sugar and almonds and scatter over the cobbler. Bake until the filling bubbles rapidly and the topping is golden brown, 45 to 65 minutes. Serve warm or at room temperature. The cobbler is best the day it is baked, but it's still pretty good the next day.

NOTE: You can omit the almond flour and use a total of 1 cup all-purpose flour instead. You'll likely need a little more milk to make the dough nice and soft.

ALSO TRY WITH: Blueberries, raspberries, nectarines, apricots, or plums.

ORANGES

Citrus spp.
Rutaceae family
Southern US in zones 9a–11

ALL FRUITS ARE PACKAGES, but oranges are a gift. Each one comes wrapped ready for you to open, or to give to someone else.

Among citruses, only the orange is ready to dig into. It is best consumed with your hands, plump segments shoved whole into the mouth, while somehow the flesh of a grapefruit improves with a more formal presentation, a plate and a spoon and a cleaved specimen cut side up. The butchery of oranges is visceral, not surgical. Have you ever made an orange do a striptease, removing a spiral of peel to see how long you could get it in one piece? Have you ever done this to impress someone you were about to share the orange with? Or perhaps it was a test of skill performed to amuse yourself, something to draw out the pleasure of the orange.

Oranges grow in places other people want to be. Only four states can grow oranges commercially: Florida, California, Texas, and Arizona. These are states that know no snow days or dragging rainy seasons. In popular thinking, the orange juice there runs like water, and the glossy leaves of the orange trees glint in the sunlight in expansive, attractive groves.

Oranges are the fruit of luxury. If a hotel aspires to be fancypants, they list freshly squeezed orange juice on their room service menu. A fruit basket is not a fruit basket without the presence of oranges, a few of them perhaps wrapped in tissue paper with some skin showing through

coquettishly. If a family relies on SNAP (food stamps, as we persistently call them), getting enough groceries to feed a household means that fresh fruit happens sporadically, maybe when oranges are on sale for 58 cents a pound at the store across town that takes two bus transfers to get to. The ride home means wrangling a granny cart full of provisions and a plastic bag heavy with oranges swinging with the movement of the bus and knocking into fellow passengers. But there is the promise that for one week, at least, there will be fresh fruit. That is the dose of luxury before scrimping for crumbs the week before the next infusion of SNAP.

The Northern California artists Pam Bolton and Cindy Cleary launched a project in 2003 they called Free Fruit / Fruita Gratis. They did installations that included giving away fruit trees to migrant workers and setting out crates of donated locally harvested fruit. The crates and the paper sacks next to them were stamped with a Free Fruit logo. Bolton and Cleary didn't know where the fruit wound up or who took it, but that's how it works to truly give something away.

Their concept touched on the attractiveness of fruit, but also its exclusivity as a product. To have a fruit tree, you have to have a yard; to buy fresh fruit, you need the store that sells fresh produce, plus transportation to said store. Oranges were once a thing not taken for granted. My mom told me she always got oranges in her Christmas stocking, a holdover from my grandparents' own hardscrabble childhoods. For those on limited incomes who reside in food deserts, oranges remain as rare as Christmas Day.

Oranges descended from mandarins and pomelos, either in southern China or Southeast Asia. Many orange-fleshed citruses (mandarins, oranges, tangerines) are an indispensable presence at Lunar New Year feasts. They symbolize gold or fortune and are preferably given with their green leaves still attached, to symbolize fertility and a long life.

They are always given in pairs, but never in sets of four, because the number four is a homophone for the Chinese word for "death."

We can divide oranges between bitter and sweet. *C. aurantium* are the bitter ones; we call them the sour, Seville, or (fittingly) bitter oranges. They are what give a Cuban *mojo* sauce its mojo.

Bitter orange juice is valued in Caribbean cooking, and makers of marmalade use it for the sort of preserves Paddington Bear ate by the jarful. For those attuned to different vices, blue Curaçao owes its flavor (not color) to bitter oranges.

Sweet oranges (*C. sinensis*) are the kind most people grow in yards on purpose, because they are easy to eat (no offense, bitter oranges). I stayed with a friend in Los Angeles one February, and his neighbor had two navel orange trees fully laden with fruit. My host assured me it was okay with the neighbor, who was not home, to nab some. "He even has this orange picker thing right by the trees. He'll be happy someone's getting the oranges. That's why he left it there." The contraption was a long handle with an open cagelike apparatus on the end. I wound up getting it stuck on a grapefruit-sized orange way up at the top of the tree; the cage remained, but the handle detached. That massive orange some 20 feet (6 m) up still retains a mythical status to me. I have thought about going back to pick it, as if it would still be there for me, years later. The oranges I did pick were wonderful, varying in size but juicy and sweet. I ate so many it tore up my mouth. We took a box up to my in-laws, and I regretted not picking more to distribute to strangers like a deranged fairy. How could anyone sane turn down such an offering, or the opportunity to bear it?

When you eat an orange, consider this: Oranges are not quickly growing fruits. A human couple can make and bear a baby faster, in some cases. To consume an orange without processing what went into it—air, sun, water, soil, time—is reckless. Collected oranges are not for hoarding. They are for giving. Listen to what those oranges are telling you and share them.

Harvesting and Storage

Oranges do not ripen off the tree. Taste one, and that'll tell you if they are ready. Just pluck the oranges from the tree in the classic fashion, or use pruning shears to cut them off at the stem. They also do not ripen all at once, giving cause to revisit the same tree multiple times.

Color is a helpful indicator, but not the main one. The full-on orange oranges we buy at the store are occasionally dyed. Don't pick green oranges, but—depending on the variety—don't hold out for evenly colored orange exteriors, either. Oranges that have fallen to the ground are fine to collect, and a probable sign of other ripe ones on the tree. Leave behind any ones with cracked rinds.

Oranges keep well at cool temperatures. It's best not to pile them up too high. You want air to circulate; this prohibits mold growth. Room temperature is just fine, though, which is just great for your Lunar New Year decorating needs. Place them around the house accordingly and snack on them or give to others (in pairs, remember, not fours!) until they are gone.

Culinary Possibilities

Do you need me to tell you what to do with oranges? You do not. If they are sweet and juicy and wonderful to eat, take what you can use and give away the rest. Don't be lazy about it, though. Don't just drop it off at a food bank sight unseen; call first and see what they can accept, and if they'd have any ideas for other places to take them if they can't. Failing that, a cardboard box with a scrawled Sharpie FREE TAKE ONE OR TEN works. Free things on the sidewalk appeal to a certain type of person, regardless of background or income. A box of free fruit is one thing that unites thrifty scrounges across ethnic, class, and language barriers.

Grocery Store Foraging

One way to be a bottom feeder besides foraging, growing, and gleaning is buying. When I shop at a market, my favorite things to get are the produce misfits. A bushel of mystery apples from a random tree on a farmer's property? "I'm not sure what they are, but they make good applesauce," the guy told me. Sold! A mesh bag of knobby lemons on the 99-cent rack at the grocery store? Sure! That's also where I get my pears—sometimes I see a bunch of quite ripe ones, and if I know I'll have time, I bring them home to make a batch of pear butter or a pear pie.

I like how these external nudges can spur creativity. The idea that we can get nearly any common food plant at our whim is ridiculous, and grocery store foraging narrows the selection down just enough to challenge us. Me living in Ohio and buying a marked-down, overripe imported papaya is not going to solve a global food supply crisis, but it's a kind of foraging I can practice when there's not a lot to harvest on the plants living under the beautiful skies outside, away from the humming lights and low-level pop music of my gently numbing grocery store.

PASSION FRUIT / MAYPOPS

Passiflora spp.
Passifloraceae family
Eastern and southern US

DISTINCTIVE AND INTRICATE, the blossom of the hearty passionflower vine understandably gets the lion's share of the attention. But the perennial climbing vine produces a fruit, one that beguiles as well, despite lacking the visual flash of the flower.

For all their exoticism, the passionflowers growing wild here are native to North America. Most passionflower vines are evergreen, but *P. incarnata*, known as maypop or apricot vine, is deciduous. It grows wild in the southeastern and central United States and specializes in appearing in otherwise mundane areas: It climbs fences near ditches, fields, and open lots where it can get full sun.

Like the pawpaw, it is a little-known delight that amazes first-timers with the tropical essence of its flavor. I, for one, cannot be the only person whose primary experience with passionfruit is via childhood cans of Hawaiian Punch at birthday parties. To eat a tangible fruit growing in a midwestern field that echoes an artificially flavored fruit drink is mind blowing.

Supposedly *maypop* derives from the popping sound of stepping on one of the fruits. Egg-sized maypops are just the right size for small feet to stomp on; children playing outside who come across a vine with maypops would probably revel in this, but more forward-thinking and less impulsive adult foragers would not consider depleting their

possible harvest this way. (Maypops do not ripen in May, by the way, but how catchy is Augustpop?)

John Muir wrote in *A Thousand-Mile Walk to the Gulf* that maypops were "the most delicious thing I have ever eaten." He was in Georgia at the time, en route from Indianapolis to Florida, and had been spotting (and presumably eating) maypops since hitting Tennessee. To crack into a shriveled fruit and discover a fragrant, pulpy, slurpable mess inside can feel like stumbling on a hidden treasure.

Maypops are not as sweet as cultivated passion fruit (*P. edulis*). The commercially grown passion fruits we see most (if we see them at all) are a dull purple when ripe, but maypops are a faint brown or yellow when ripe. *P. caerulea* grows in the Southeast and West and has blue flowers and fruits that ripen to a yellow-orange. Its fruit is edible, but not as enticing as other species. It's one of the most widely cultivated ornamental varieties. All *Passiflora* do well with full exposure to the sun.

The skin and pith are technically edible, but bitter. All parts of the plant have a medicinal application as a sedative. Passionflower shows up in the homeopathic sleep aid tablets I take sometimes. I'm not sure if they work because they truly work, or just because I believe they do.

Passion fruit vines will spread with aplomb, if allowed to. When not growing in the wild, they can be found climbing up backyard trellises. Typically in such cases, the passionflowers are grown for ornamental purposes, and the fruit is secondary, if utilized at all.

The anatomy of the flowers has been elaborately shaped into weighty Christian symbolism of the Passion of Christ. The five petals and five sepals total ten, representing the ten disciples who remained faithful to Jesus (there's Peter, who doubted, and Judas, who sold him out). The radiating curly purple filaments, called the corona, represent the crown of thorns. The five stamens stand for the limbs and head of Christ as he was nailed on the cross, and the three stigmas of the pistil represent the three nails that created the stigmata. The vines stand for the whips the Romans used.

To assign such gory attributes to a flower of such ethereal beauty is the whole crux of the biscuit: From passion comes grace. Thomas Johnson, editor of the 1633 edition of *Gerarde's Herbal*, took a less romantic view. "The Spanish Friers for some imaginarie resemblances in the floure, first called it Flos Passionis, The Passion floure, and in a counterfeit figure, by adding what was wanting, they made it as it were an Epitome of our Saviors passion. Thus superstitious persons semper sibi somnia fingunt." That last Latin bit means "always see contrived

FUCHSIA

Fuchsia spp.
Onagraceae family
Zones 10–11

People grow fuchsias for the flowers. Consider the fruit a bonus.

The first weekend in May, I buy a hanging fuchsia plant from a local greenhouse in anticipation of summer's arrival. The variety I buy was bred for its supersaturated pink-and-purple flowers. It is a celebratory purchase, and immediately I hang it out front to initiate the season of good porch weather. I enjoy a lot of solitary meals on the front porch, and I think of the fuchsia as my companion during the months I love best. The hummingbirds dig it, too.

To many people outside of tropical zones, fuchsia are colorful container plants, but they grow as giant shrubs in warm regions all over the world. There are many species, and more than three-quarters of them can be found in South America. My dad once visited Bolivia and brought back two pink-gold pins bearing what we assumed were dangly narrow bell shapes; only recently did we discover they depicted fuchsia blossoms.

Even my porch fuchsia produces a succession of small berries, starting around July. They are dark, smooth, and oblong; the softer ones are the ripest and have the most flavor. They are mild, sweet, and juicy—overall, pretty good little nibbles. I probably only eat five a year and am not compelled to eat more; their finish is faintly vegetal. Perhaps if

I lived in a region where it was possible to gather buckets of fuchsia berries, I'd be more taken with them. I even ran across a fuchsia jelly recipe, but most of us are about a hundred plants away from having enough berries to make that a reality.

The plant's name honors the sixteenth-century German herbalist and professor Leonhart Fuchs, who incidentally never laid eyes on one. It was named by a French monk, botanist, and Fuchs fan, Father Charles Plumier. In 1695, while visiting what is today the Dominican Republic, Plumier encountered the fuchsia and introduced it to Europe (along with lobelia, begonia, magnolia, and other now-familiar plants). If, like me, you have always had trouble spelling *fuchsia*, remember the Fuchs connection.

images." We see what we want to see. Passionflower grows with no inkling of the symbolism placed upon it, liberating you to form your own folklore around its configuration.

Harvesting and Storage

Wrinkled but not overripe fruits are what you want to look for; smooth-skinned passion fruits need to ripen longer. Their color will depend on the species, but they should feel heavy for their size. You can keep passion fruits in a cool, dry spot for up to a few weeks.

Pop into the skin and suck out the seeds, which are covered in slimy arils. You can eat the seeds, which are crunchy and soft, or spit them out. Freezing the juice or pulp is a possibility, if you have more fruits than you can use.

Culinary Possibilities

To keep the flavor of ripe specimens nice and fresh, eat them raw. Cut them in half, and use a spoon, if you like. The simplest way is to top ice cream, cereal, or fruit salads with the seedy pulp. Run the pulp in a blender (straining is optional) to get the juice, which you can add to smoothies or use as a base for sorbet. Ripe or not-quite-ripe passion fruit can be cooked into jam or jelly.

Because passion fruit is especially intense and aromatic, it is often paired with other fruits, but if you find ripe ones ready to eat while tromping around one day, ripping one apart right then and there is recommended. That's the magic of ripe fruit tasting better where you pick it.

Mortality and Fruit

After you observe the same plants over a series of seasons, the cycles of life and birth embed themselves deeply in your spirit. The weather turns, leaves wither, and any fruit not utilized by humans or animals shrivels or decays. Then come the months of passing through familiar routes in their bleak, barren state. It can make them look unfamiliar. Some plants utterly disappear; some remain, skeletal and scratchy. I'll seek out my favorite groves of pawpaws or black raspberry brambles and wonder if anything ever grew there at all. Do pawpaws exist? Did I ever love them? The sensations of holding them and smelling them grow hazy in my memory.

Yet I still go out there, checking in for signs of life. Obviously I know the plants will return to their summer splendor, but in the middle of winter it can seem beyond distant, like another world. The cycles are gradual, though, and not a bummer. What would be worse is no change at all.

What is the opposite of mortality—immortality? We think of mortality applying to humans only—to us mortals—but mortality means being subject to death, and every living thing will eventually die. There is no life without death. I like living things, and I like being alive, and the inevitability of everything I love dying is part of that experience. I simultaneously mourn and rejoice in the mortality of the fruit plants I hold so dear, and these small doses help me come to terms with our human mortality. What's an immortal fruit? A fake fruit, as fake as the plastic cucumber I bought at a garage sale because it reminds me of a penis and cracks me up (thank you, Spinal Tap). Our gentlemen readers might perk up at the concept of having an immortal penis, but to me it means a fake penis. Isn't a mortal penis so much more welcoming?

Generally, we can't pick out exactly how or when we die. I don't want to die tomorrow, but I could, which prompts me to consider the more meaningful parts of the life I've had so far. I've lived in a lot of different neighborhoods, in massively varying states of contentment. Even when things were going badly, some of my fondest memories of any given time are of explorations on foot along otherwise dingy sidewalks or boring roadsides, and of the centering power that being outside my typical realms gave me. Our entire lives are transitions, just as walking is a transition—continuous movement to a destination. Maybe that's the ultimate source of my obsession with fruit watching: It's a self-soothing reminder that I, too, will meet the same fate as rotting pears and desiccated crab apples. I watch the fruit be mortal, and it eases the sting, because I'm not alone.

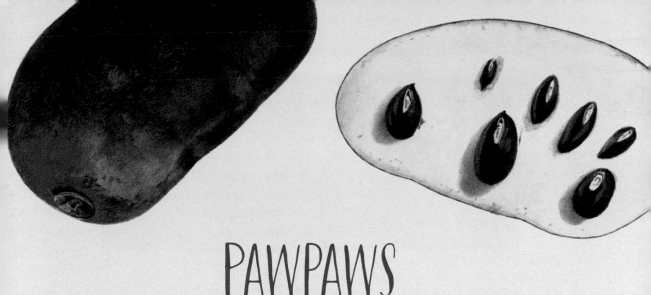

PAWPAWS

Asimina triloba
Annonaceae family
Eastern and southeastern US; southern Ontario

A PAWPAW IS LIKE NOTHING ELSE. Its scent, its texture, its form—a pawpaw confounds the senses and defies expectations, in the most enchanting of ways. Pawpaws have grown in North America for some fifty-six million years, and today they promise unknown possibilities. They taste of sunshine and fragrance and a good fruit salad: Mango and banana are the common touchstones, though citrus and pineapple are in there, too. Hoosier banana, Indiana banana, false banana, and hillbilly banana are all nicknames the pawpaw has well earned.

Pawpaw trees grow wild in deciduous forests in a huge swath over eastern North America, from southern Ontario to northern Florida and as far west as eastern Kansas. Many people in rural areas grew up eating them out in the woods. On the fringes of Appalachia, where I live, pawpaws are a folk fruit. To people of that older generation, they call to mind the freedom of playing outside as a child and the thrill of getting their hands all over nature. "My papaw had pawpaw trees on the woods near his farm," they'll say in a wistful tone.

Those traditions were often lost in the age of urban sprawl and supermarkets. I'm a prime example: Pawpaw trees thrive in the woods behind my childhood home, but I never heard of the fruit until I was an adult. It's worth noting the trees behind my parents' house never bear fruit. That's not unusual. You can run across sizable stands of

pawpaw trees growing in a forest, yet they just don't fruit. Or some might fruit well one year and not the next. Pawpaws are tricky that way, because they can be. Wild trees send out root suckers and often form colonies without spending a single seed.

Spotting the fruit itself, like most foraging, gets easier with time. Pawpaws are egg-shaped and have speckled green skin that does not exactly pop out at you when you look up into the leafy canopy. In early spring the trees produce small brown flowers that reportedly smell of carrion. I have sniffed dozens of pawpaw blossoms and not found this to be the case, but maybe my senses are off. Besides, that reputed stench is not there to attract me—it's to attract insects to pollinate it.

The leaves are large, long, and tapered; the trees are spindly. They can grow tall—up to 40 feet (12 m)—but are usually under 12 feet (3.7 m). The fruits begin to appear in the early summer and can form on the branches singularly or in clusters of two to five or even seven. (Every now and then, you might run across a specimen that appears to be two small pawpaws fused together and resembles a smooth, prettied-up scrotum.)

Pawpaws are the largest fruit native to North America, but a wild pawpaw big enough to fill your palm is, in pawpaw terms, huge. Cultivated pawpaws typically bear larger fruits because that's what those cultivars were bred to do. Global interest in pawpaws as a commercial crop has picked up in the last few decades, but there are some hurdles in harvesting, processing, and marketing to be cleared first.

If you pick up a ripe pawpaw, you'll know why. They are smooshably fragile. It's hard to imagine them lasting long in a produce aisle, and once you get into one, it turns to a creamy mush in seconds. If you can dice a pawpaw, it's not ripe.

With pawpaws the party is inside. The skin is not the part you eat. The flesh is impossibly fragrant and unbelievably tropical. Some have pale flesh, while others are a bright orange-yellow. And each one has

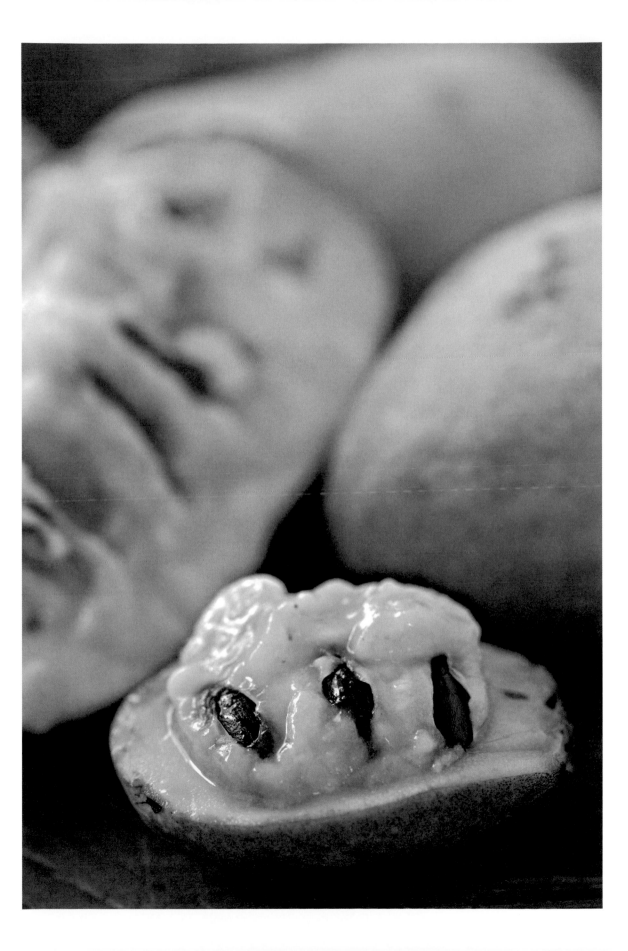

tons of almond-sized seeds that look to be black, but are actually dark brown and glossy. Spit out the seeds, which are prone to cause digestive discomfort if you bite into one and swallow it. Besides, they don't taste good. Critters in the woods, meanwhile, slurp down pawpaws whole, as evidenced by telltale scat dotted with plentiful pawpaw seeds.

It's fun to tear into a ripe pawpaw on the spot like a savage wolf-person, flinging aside the skins and wiping your gooey hands on your pants. For more refined use, processing the pawpaw is required to remove the skin and seeds. What you're left with will look like runny baby food. That's the pulp. It is not literally worth its weight in gold—oh, if only!—but it is a treasure with immense potential.

Harvesting and Storage

Pick pawpaws only when they are ripe; they don't do much ripening off the tree to speak of. If you spot a pawpaw—particularly a larger one—look at its underside: When some are wonderfully ripe, they'll have a faint, powdery bloom and a hint of a yellowish aura. I call this the pawpaw glow. The technical term is *color break*, but it's more than a color; it's an energy. Not all pawpaws have it, but it's a sign you are on the right path.

Do get handsy with pawpaws, but don't grope them, because they bruise easily. Ripe pawpaws will give way slightly to gentle pressure from your thumb, and they will separate from the branch with barely any resistance. For pawpaws that are too high to reach, try swatting at the branch with a stick or shaking the tree firmly but gently. If the pawpaws are ripe, they'll come tumbling to the ground.

In a spot rife with fruiting pawpaw trees, you'll likely see pawpaws on the ground in assorted stages of ripeness/decay. I do pick up the ones that are in good shape—they may have fallen quite recently—but if they are ruptured, squishy, or brown, then leave them behind.

Getting pawpaws out of the woods unharmed is no small feat. I always have a few casualties, even if I don't pile them up on one another. You have to just accept it. I've used small canvas sacks, a backpack, plastic bags, egg cartons—you name it. (Pawpaw goop will oxidize and stain fabric, by the way, so either use a bag you don't care about or wash it in a timely manner.) Sometimes I get too carried away and lose track of time and realize I need to hustle out of the Pawpaw World and back into my own. It is what it is, but even hiking out at a brisk pace can jostle these things.

Very soon after returning home, sort through your pawpaws. Refrigerate or use any damaged ones immediately. The rest can hang out at a cool room temperature for a few days, or for up to a week in the refrigerator. I feel their flavor is usually better if you deal with them sooner than later, but a small percentage of pawpaw fans like to let theirs get very soft and brown, like nasty old bananas. This is a personal preference thing, though I call that state "rotten."

If you want to preserve pawpaws for later, freezing is the way to go. I use the pulp and like to have it recipe-ready, so I process it the day I harvest. It's incredibly hands-on. You'll need a conical food mill or colander. Rinse off the pawpaws, then dig your thumbs into the pawpaw like you are cracking an egg, and pry it in half. Cup one pawpaw half in each hand, and use your thumbs to scoop the flesh and seeds out into the food mill. Discard the skin, which is flimsy enough that some bits may make their way into the flesh; try to get those out. Then force the flesh through the mill with a pestle, or through the colander with your hands. Really work it. I start out using my pestle and then switch to my hands, which helps me get a lot of pulp fast. I've talked to people who painstakingly remove the last stubborn bit of pulp that clings to each seed like a slimy jacket, but I think that wastes time.

Pawpaw pulp oxidizes like the dickens. To quell this, add about a teaspoon of lemon or lime juice for every cup, or use some ascorbic acid. Then measure your pulp, and put it in heavy-duty zip-top freezer bags. I like to freeze the pulp in 2- or 4-cup (480 or 960 ml) parcels.

If you're pressed for time, you can skip the pulping and just freeze the skinless seeds and flesh. Or even pop entire pawpaws in the freezer.

I feel pawpaws are ill suited to canning because their flavor is so sensitive to heat. Unless you are up for bold experiments, skip dehydrating them—I have heard of people consuming pawpaw fruit leather and feeling nauseated. And one more word to the wise: Composting pawpaw seeds will give them the exact conditions they need to overwinter, ferment, and germinate happily in any soil you reap from your composting. That may seem cool, but pawpaw seedling taproots go down very deep. I pulled at least forty from my small raised bed the year after I was first struck with pawpaw mania.

Culinary Possibilities

Repeat after me: Pawpaws taste best when exposed to the smallest amount of heat possible. Ideally that's zero. Many pawpaw lovers eat them raw and unadorned and no other way. *Pawpaw* author Andrew Moore likes to halve his and work his way through with a spoon. Plopping some into a smoothie is also a shining, low-entry use for pawpaws.

Rain

Don't let rain keep you indoors. Some of my happiest foraging outings have been under drizzly skies.

Delicate berries like raspberries and currants are hopeless to harvest when it's wet out, because they'll quickly turn to mush. But other fruits are easier to gather after (or even during) a spell of rain. Raindrops sometimes knock ripe pawpaws from their branches, and if the stars align just right there will be a pawpaw bonanza to collect on the soft, leaf-lined ground below. The best time to collect American persimmons (which you collect, never pick) is after a rain.

Things look greener when it's raining. They *smell* greener. A leafy summertime woods becomes a jungle when rain clouds are overhead. There's a completely different sense of place, a more primeval vibe than you usually get—even in a residential neighborhood. Suit up in knock-around clothes, get out there, and get muddy. Dry yourself and your fruit off afterward, and you'll savor the comfort of shelter all the more, with some new insights to boot.

As much as I love pawpaws, they are so dang intense I have a hard time processing their flavor unless it's concocted into something else. For sweet flavors, warming spices, citrus, dairy, and vanilla are all good matches. For savory applications, bright and forceful pairings work best: mustard, habanero peppers, barbecue sauce.

The workaround to the no-heat rule is to use pawpaw as a finishing ingredient. Chef Dave Rudie finishes warm savory sauces, like beurre blanc, with a dollop of pawpaw. Add pawpaw to a cooked ice cream base *after* it has cooled.

One exception I've noticed to the no-heat rule is high-moisture, high-sugar baked recipes; I think the sugar helps to cloak any flavor changes, and the moisture provides a buffer. Custardy baked pawpaw puddings or crème brûlées are divine.

The Albany, Ohio, pawpaw grower and processor Integration Acres makes a line of pawpaw preserves and chutneys. They pull this off by combining pawpaws with high-pectin fruits, like apples, to keep the funkiness at bay and allow the preserves to set. I once made pawpaw preserves with no pectin, and it turned out like a giant batch of quixotic fruit butter. The only thing it tasted good with was chicken liver pâté. It took me years, as well as plenty of forced enthusiasm, to work my way through.

Pawpaw Gelato

Makes 1 scant quart (960 ml), to serve 6–10

Generally speaking, the difference between gelato and ice cream is air and fat. Ice cream has more air churned into it and a higher butterfat content. Gelato? Less air, less fat. But no less enjoyable, and so incredibly smooth.

1 cup plus 3 tablespoons (285 ml)
   half-and-half, divided
2 tablespoons cornstarch
⅔ cup (130 g) granulated sugar
½ cup (120 ml) sour cream
   (preferably full fat)
2 cups (480 ml) pawpaw pulp
1 tablespoon vanilla extract
2 teaspoons lemon juice

Combine 3 tablespoons (45 ml) of the half-and-half with the cornstarch in a small bowl. Stir until smooth; set aside.

In a medium saucepan, bring the remaining cup (240 ml) of half-and-half and the sugar to a boil over medium-high heat (keep an eye on it, as the mixture can boil over easily). Reduce the heat, and keep at a robust simmer for 4 minutes. Stir the cornstarch mixture once more, and gradually whisk it into the boiling half-and-half. Once it returns to a boil, cook for 1 full minute, whisking all the while; it will become thicker, but it still won't be terribly thick. Scrape into a large bowl, and whisk in the sour cream. Let the mixture cool completely (I recommend using an ice bath).

Once cool, add the pawpaw pulp, vanilla, and lemon juice. For a silky-smooth gelato, puree the gelato base using a blender, food processor, or immersion blender for 1 minute.

Churn the gelato base in an ice cream maker according to the manufacturer's directions. The gelato will have the consistency of a thick, stiff milk shake when it's ready. Pour it into a shallow metal pan (it cools the gelato faster and will be easier to scoop from), and place in the freezer for at least a few hours to set up.

The frozen gelato will keep, covered with foil pressed directly on its surface, for about 1 week. For optimal enjoyment, let it soften for about 10 minutes before serving.

PAWPAW PUDDING

Serves 8–12

Smooth and rich with intriguing caramel notes, this old-fashioned pudding has an undeniable pawpaw kick. With a food processor, it takes only minutes to blitz that batter together. (Note: Minutes blitzing together batter excludes gathering of pawpaws.) Reserve a stash of frozen pawpaw pulp to make this at Thanksgiving. It'll blow pumpkin pie right out of the water.

⅔ cup (85 g) unbleached
    all-purpose flour
⅔–¾ cup (130–150 g) granulated sugar
    (I prefer a less sweet pudding)
¼ teaspoon table salt
½ teaspoon baking soda
1 egg
1 egg yolk
1 cup (240 ml) pawpaw pulp
½ cup (120 ml) buttermilk,
    preferably whole-milk
¼ cup (60 ml) half-and-half
2 teaspoons vanilla bean paste
    or vanilla extract
3 tablespoons unsalted butter,
    melted and cooled

Preheat the oven to 350°F (175°C), and position a rack in the middle. Grease a 9 × 9-inch (23 × 23 cm) baking dish, preferably glass or ceramic.

In the bowl of a food processor, pulse the flour, sugar, salt, and baking soda to combine.

In a large glass measuring cup or medium bowl, combine the egg, egg yolk, pawpaw, buttermilk, half-and-half, and vanilla bean paste. With the food processor running, add the pawpaw-buttermilk mixture through the feed tube. Turn off the machine, scrape down the sides, and add the butter with the machine running. Your batter should have the consistency of pancake batter.

Pour the batter into the greased dish. Bake until the center is set but still jiggly (like a pumpkin pie), 30 to 45 minutes. The sides of the pudding will rise up and brown, while the interior will be flat, shiny, and amber-colored. Let cool to room temperature, and serve with crème fraîche or whipped cream.

The pudding will keep 2 to 3 days at room temperature. I suppose you could refrigerate it, but I like it better when it's not cold.

Pawpaw Ha-Ha Habanero Hot Sauce

Makes 3–3½ cups (720–840 ml),
depending on how many habaneros you use

Bright orange and hot as hell, this sauce was inspired by the Ha-Ha Habanero sauce at Laughing Planet, a cheeky bourgeois burrito chain in Portland, Oregon. My husband used to sneak out tiny plastic to-go cups of the stuff.

Putting 10 of them in this sauce is a little bonkers. You can use as few as two, if you are more reasonable. Over time, the hot sauce will mellow out, and the other flavors will come through more.

1 teaspoon vegetable oil

1 onion, chopped

5 cloves garlic, peeled

2 carrots, grated (enough to make
    1 packed cup)

¼ cup (60 ml) lime juice

¼ cup (60 ml) white wine vinegar

2 tablespoons honey

As few as 2 and as many as
    10 habanero peppers, halved,
    seeded, and stemmed (if possible,
    wear disposable gloves when you
    handle them)

1 teaspoon kosher salt

1 cup (240 ml) pawpaw pulp

Heat the oil in a medium saucepan, and add the onion. Cook over medium heat until soft. Add the garlic and carrots and enough water to barely cover the carrots. Bring to a simmer, and cook for 10 minutes.

Transfer to a blender or food processor, and add the lime juice, vinegar, honey, habaneros, and salt. Puree, adding a little water if it's too thick. Cool until tepid, then add the pawpaw pulp, and puree again. Cool completely before putting in bottles or jars. Store tightly covered in the refrigerator for up to 2 months, if not longer.

NOTE: All year long I wash out empty sriracha bottles and save them so I can make a batch of this come pawpaw season and give bottles away to my heat-loving friends. It's pretty impossible to enjoy unless you can easily dole it out in small drops. Wear gloves when handling habaneros, and refrain from touching your eyes and face to avoid irritation.

PAPAYA

Carica papaya
Caricaceae family
Southern US in zones 10–12

Papayas are not pawpaws, and pawpaws are not papayas. In Australia and some Spanish-speaking countries, *pawpaw* is another word for "papaya," so depending on where you are, a papaya *is* a pawpaw. But the plants are not related, and the pawpaws the carefree bear Baloo sang about in Disney's adaptation of *The Jungle Book* are papayas, not *Asimina triloba*.

Both taste great, though, and have very soft flesh when ripe. A friend in Miami has a papaya tree in her yard—the southeastern United States is where you'd find them growing happily as ornamentals, though the trees are not long-lived. Papayas are thought to be native to Mexico and Central America. They like warm to hot weather all year long, detest frost, and can grow well in containers or greenhouses with a doting caretaker.

Many of the papayas we see at the grocery store are the size of small winter squash, but some can get really honking big. The preference is to pick them before they are ripe; they'll get good and ripe on a counter over time. In fact, some people like to let them get to the point where they are soft and spotty and almost rotten looking. Fans of Southeast Asian cuisine are also acquainted with unripe papayas in the form of *som tum*, the popular green papaya salad of Thailand. Green papayas are firm and bland, and when shredded, dressed in lime juice, and tossed with dried shrimp, fish sauce, chilies, and peanuts, they make for a completely different experience from eating a soft and juicy ripe papaya.

Pawpaw Lemon Curd

Makes 3½ cups (840 ml)

Pawpaw pulp, which is custardy to begin with, works wonderfully with fruit curd. It's the same base recipe as Citrus Curd (page 192), but omits the zest so it doesn't overpower the pawpaw. You blend the cooled curd with pawpaw pulp, and it becomes its own marvelous beast. If you don't anticipate using all of the pawpaw curd within a month, simply freeze half to use later.

2 large eggs

2 large egg yolks

¾ cup (150 g) granulated sugar

½ cup (120 ml) freshly squeezed lemon juice, strained

6 tablespoons unsalted butter

2 cups (480 ml) pawpaw pulp

1 tablespoon vanilla extract

In a medium stainless-steel, nonstick, or enameled saucepan, beat the eggs, yolks, and sugar with a whisk until the sugar is mostly dissolved. Whisk in the lemon juice.

Place the pan over medium heat, and cook, stirring constantly with a heat-resistant spatula, making sure to scrape the bottom and corners of the pan. Slowly, the mixture will turn more opaque and the spatula will start to make visible swaths through the mixture, 10 to 15 minutes. Keep stirring until the curd is as thick as sour cream and coats the spatula, 2 to 3 minutes more.

Remove the pan from the heat, and stir in the butter until it melts and the curd is smooth. Pour it into a medium bowl (there's no need to strain this, since it gets pureed later). Lay a piece of plastic wrap directly on the surface of the lemon curd to prevent a crust from forming. Chill in the refrigerator until cool and set, at least 4 hours.

Remove the lemon curd from the refrigerator. Using an immersion blender, food processor, or blender, puree the pawpaw pulp, vanilla, and cooked curd together until smooth and combined. The curd will be creamy and opaque. Transfer to a clean glass jar, and cover tightly. The curd will keep for about a month in the refrigerator, or in the freezer for up to 1 year.

PEACHES

Prunus persica
Rosaceae family
Throughout the US and Canada in zones 3–9

PEACHES ARE OF THE NOW. An underripe peach is cruel in its refusal to let up any juice or yield to a paring knife or tooth. It is a promise unfulfilled. Ripe peaches seduce, begging to be eaten messily at the very moment a hand graces their velvety skin.

Singer-songwriter Laura Veirs's song "July Flame" best articulates the poignancy of peaches:

> *July Flame*
> *Sweet summer peach*
> *High up in the branch*
> *Just out of my reach*

The song then goes into a refrain of "Can I call you mine?" That's the sting with anything wonderful. You can only call it yours in that moment: the precise way dry leaves crunch under your feet in the fall, the buoyant walk of your first grader heading across the playground to school in the morning, the fluttering of peach blossom petals cascading downward after a spring breeze.

Peaches are rewarding trees even when they do not bother to produce peaches. Frilly and pink or white, the blossoms appear before the slender, tapered leaves do. The trees have a sturdy but more

feminine vibe than other Rosaceae, like apples or cherries. Early on, Europeans believed that peaches were native to Persia (their scientific name means "Persian plum"), though research suggests they were first domesticated in China's lower Yangtze Valley around 8,000 BCE. Peaches did arrive in Europe via Persia.

Spanish explorers planted peaches in the New World in the sixteenth century, and escapees in the southeastern and mid-Atlantic colonies were prevalent enough that explorer and naturalist John Lawson wrote, "We are forced to take a great deal of Care to weed them out, otherwise they make our Land a Wilderness of Peach-Trees." Some explorers even assumed it was a native tree.

A wilderness of peach trees does not sound like too bad a fate. Thomas Jefferson had his staff at Monticello plant nine hundred of them, the opposite of wilderness, creating dividing lines between his fields and those of adjacent property. What to do with the fruit of nine

hundred peach trees is its own matter, though Jefferson mentioned in a letter to his daughter that his slaves dried some of them.

The gift of a windfall of peaches depends on the state of the peaches. I got a box of small, hard ones from a friend of a friend and had to get creative. They were too firm to eat raw, and too scant in flavor to justify peeling them painstakingly. I wound up rinsing them off and simmering them in water to make a pulp that I cooked down into peach butter. Dressed up with brown sugar and almond extract, it turned out pretty well—though just imagine the flavor if those little guys had been ripe!

Peaches are either clingstone or freestone. Clingstone peaches are usually the first to be ready to harvest. As their name suggests, their flesh clings to their pits, no matter how much finesse you put into twisting two halves apart. Well, then, what's the point, you say? If you like that particular variety enough, you'll just deal with it. Many commercially canned peach products come from cling peaches. I'd love to know what their secret is for getting the dang things off the pits without mangling them or wasting half the flesh. I sure haven't.

Freestone peaches, when you score the fruit along its peach butt crack, will slide right off their pits. Naturally these are the preferred peach for all kinds of purposes. Peach grower Ben Wenk of Three Springs Fruit Farm in Pennsylvania says there's no default superiority in the flavor of either. "If one peach tastes better than another, it's because the variety is good, not necessarily because it fell off the pit."

There is an actual July Flame peach, part of a series that also includes Summer Flame and August Flame. Peaches are of summer and the promise of more hot and sunlit months to come, even though the peaches won't be there for all of it. The arrangement is cyclical but temporary. *Can I call you mine?* Every time you walk up to a fruiting plant, isn't that what you're really asking it? And every time, I'm still not sure of their answer.

Harvesting and Storage

Pick peaches ripe or just shy of ripe. Heaping them up will bruise the bottom layer, so seek out shallow and flat containers, like plastic dishpans or sturdy cardboard boxes. When you get home, sort through the peaches, separate any mushy or damaged ones, and cut those up within a day, or they'll rot even more.

Lay underripe peaches on newspaper in a single layer in a box, or put them in paper bags. They'll get softer, and their flavor will improve after a few days, but they'll never live up to a peach totally ripened on the tree.

Peaches are simple to can or freeze, and they hold up to either method splendidly. Score their skins and blanch them in boiling water, then in an ice bath, and their skins should slip right off. Treat any peaches you'd like to slice and dehydrate in an ascorbic acid solution.

Culinary Possibilities

Cobbler, pie, cereal toppers, smoothies, jam, ice cream—I assume you don't need my guidance. Use raw peaches in fruit salsas and relishes to break out of the sweet rut. Halve them and mark them on a grill to serve with desserts or dinner. Add them to tangy sauces, like barbecue, for a change of pace from tomatoes.

HONORABLE
MENTION
★ ★ ★ ★ ☆

NECTARINES

Prunus persica
Rosaceae family
Throughout the US and Canada in zones 3–9

The difference between peaches and nectarines in botanical terms is the fuzz: Peaches have it, nectarines don't. Nectarines are simply a type of peach. The trees can be less hardy and more challenging to grow. You can use the two interchangeably, though nectarines have the advantage of no mouth-clogging flocked skins.

Peachy Barbecue Sauce

Makes about 1½ cups (360 ml)

If you have a lot of blemished, bruised peaches and you're not into baking desserts or making jam, try this. You can taste the peaches, but they don't dominate, meaning it actually tastes like barbecue sauce and not a gob of peach jam. It's good on pork, chicken, salmon, or (my favorite) thin slices of smoked tofu piled on a bun with coleslaw.

1 pound (455 g) peeled, pitted, and
    roughly chopped peaches
⅓ cup (80 ml) apple cider vinegar
¼ cup (55 g) packed dark brown sugar
1 tablespoon Worcestershire sauce
2 tablespoons molasses
2 teaspoons ground (dry) mustard
¼ teaspoon cayenne pepper
¼ teaspoon chipotle powder
1 teaspoon onion powder
1 teaspoon garlic powder
2 teaspoons smoked paprika
½ teaspoon freshly ground
    black pepper
½ teaspoon kosher salt

Combine all of the ingredients in a medium saucepan. Bring to a boil, stirring constantly, and reduce the heat to a simmer. Cover, and cook 10 minutes, stirring from time to time. Uncover, and continue to cook, stirring often and gradually reducing the heat until the sauce is reduced by half, 40 to 50 minutes. The sauce should be thick, and the peaches should be falling apart.

Puree the sauce with an immersion blender until it reaches the desired consistency (you may like it chunky, you may like it smooth). Taste and adjust the seasonings, if necessary. Cool and transfer to a jar. The sauce will keep, refrigerated, for about 2 weeks.

NOTE: You could double this and can it. Ladle the hot, cooked sauce into sterilized half-pint jars and leave ½ inch (1.3 cm) headspace. Process in a boiling-water bath for 20 minutes.

ALSO TRY WITH: Ripe nectarines. Keep the peels on if they are nice and tender.

1-1-1 Peach Cobbler

Serves 8–12

This straightforward cobbler has been around for years and years. My family used to vacation in a rented beach house every summer on Edisto Island, South Carolina, and this was always on the menu. It's like a very dense cake, more typical of southern cobblers than biscuit-topped ones. The name *1-1-1* refers to 1 stick of butter, 1 cup of milk, 1 cup of flour, and 1 cup of sugar (correct, that's actually four 1s). The traditional method is to dump the fruit and batter over the hot butter in the baking pan, which initially looks like it'll be a disaster, but eventually it bakes into an adorably homely dessert.

½ cup (1 stick) unsalted butter

1 cup (130 g) unbleached all-purpose flour

1 cup (200 g) granulated sugar

1½ teaspoons baking powder

¼ teaspoon table salt

1 cup (240 ml) milk

8 cups (1.5 kg) pitted, peeled, and sliced peaches

Preheat the oven to 350°F (175°C). Position a rack in the center.

Put the butter in a 13 × 9-inch (33 × 23 cm) baking dish. Put the dish in the oven until the butter melts and gets so hot it's sputtering a bit.

While you wait for the butter to melt, whisk the flour, sugar, baking powder, and salt together in a medium bowl. Add the milk, and stir to combine. The batter will be like thick pancake batter, and a few small lumps are okay.

Scrape the batter over the hot butter, then scatter the fruit on top. Bake until a toothpick or wooden skewer inserted in the center comes out clean, about 60 minutes (tent the cobbler loosely with foil if the edges start to brown too much before the center is fully baked). Cool on a wire rack for at least 15 minutes. Serve warm or at room temperature. The cobbler is best enjoyed the day it is baked, unless you eat it for breakfast the following morning. At room temperature, the cobbler will keep for 2 days.

ALSO TRY WITH: Nectarines or sweet and juicy plums. In either case, you can leave the skins on.

To Peel or Not to Peel

Sometimes fruit peels are tough and gross, or warty and blemished, and you don't want them in your food for textural or visual reasons. But increasingly, I leave on peels when I can, thanks to Sean Rembold, a New York City chef who told me how at one restaurant where he worked, the kitchen was so small that "out of necessity, we were more willing to let food be what it was. We were getting really great peaches, so we didn't peel them. If you can eat the peel, it might as well go in there."

You don't need to worry about chemical residue with fruit you forage from unsprayed trees and shrubs. Apples, peaches, and nectarines are all among the conventionally raised produce items with the most pesticide residue. When I buy them from the grocery store, I peel them. But with unsprayed, thin-skinned fruits, save yourself some time when it makes sense, and let the food be what it is. Fruit butters, purees, smoothies, and even some cobblers and pies can benefit from a little more skin now and then.

ORCHARD SALSA

Makes 2 cups (480 ml)

The sophisticated heat of habanero peppers complements the succulent ripe stone fruit in this tangy-sweet salsa. It's a wonderful destination for blemished fruit that's a little soft, but not over the hill. Try this on grilled chicken, salmon, tuna, or shrimp. It's great on the obvious: tortilla chips. The Athens, Ohio, restaurant Casa Nueva Cantina serves this, taking advantage of whatever fruit is in season at the time.

½ cup (75 g) finely diced red onion

2 tablespoons roughly chopped fresh cilantro, stems included

2 tablespoons plus 1 teaspoon freshly squeezed lime juice

¼ teaspoon cumin

¼ teaspoon fine sea salt

¼ habanero pepper, seeded, ribs removed, and minced

2 pounds (910 g) ripe peaches or nectarines, pitted and cut in ½-inch (1.3 cm) chunks (2 cups/370 g prepped)

Combine all of the ingredients in a food processor, and pulse until chunky but not pureed. Dip a chip in the salsa, and taste to check for seasoning, then serve. Salsa will keep, covered and refrigerated, for about 3 days.

NOTE: Unless the skins are especially leathery or fuzzy, there's no need to peel the fruit. You could use a serrano or jalapeño pepper instead, if you like.

Wear gloves when handling habaneros, and refrain from touching your eyes and face to avoid irritation.

Stone Fruit Sherbet

Makes about 3¼ cups (780 ml), enough to serve 6

"Sherbet?" my husband exclaimed. "I've never heard of sherbet. Do you mean sher*bert*?" In America, sherbet/sherbert is an icy, somewhat creamy frozen concoction of citrus and fruit with a little dairy thrown in for good measure. As a kid, I always thought of it as a vastly inferior stand-in for real ice cream: better than nothing, but certainly not a first choice.

Homemade sherbet is a revelation when it's packed with ripe fruit and not bobbing in a giant punchbowl filled with ginger ale and God knows what else. Tart and sunny, this is as much about lemons as it is about stone fruit, and it couldn't be easier to make. "Thanks for the delicious sherbet," my husband said. Maybe it does sound better with the extra *r*.

1 pound (455 g) ripe peaches
    or nectarines
½ cup (120 ml) cold water
¼ cup (60 ml) lemon juice
1 cup (240 ml) half-and-half
½ cup plus 2 tablespoons (125 g)
    granulated sugar, divided

Peel, pit, and roughly chop the fruit. You should have about 3 cups.

In a blender or food processor, combine all of the ingredients, and blend until very smooth. Chill in the refrigerator until cold.

Transfer to an ice cream maker, and freeze according to manufacturer's directions. The sherbet is ready to remove from the ice cream maker when it's the consistency of a Wendy's Frosty: thick but not stiff. Scrape the sherbet into a shallow pan for easier scooping (I like to use a standard-sized metal loaf pan). Cover with plastic wrap, and firm up in the freezer for several hours before serving. Let any leftover sherbet soften up on the counter for 10 minutes or so before scooping.

NOTE: You can use heavy cream instead of half-and-half, but the sherbet won't be as light and refreshing.

HERBED WHITE SANGRIA

Makes enough for a party, maybe 12–16 servings

Why make a small pitcher of sangria? When you mix lots of fruit and booze and wine together, you will want to ask all of your favorite people to drink it with you.

1 large or 2 small oranges, halved pole to pole and sliced very thinly crosswise

1 lime, halved pole to pole and sliced very thinly crosswise

¼–½ cup (50–100 g) granulated sugar

1 cup (240 ml) quality orange liqueur (such as Grand Marnier or Cointreau)

2 ripe but firm peaches, peeled and sliced

2 cups (225 g) sweet cherries, pitted and halved

4 large sprigs fresh dill

4 large sprigs lemon balm or mint (add more if using lemon balm, less for mint)

8 large basil leaves, or 3 sprigs basil

2 (750 ml) bottles dry white wine (inexpensive Sauvignon Blanc or Pinot Grigio is good)

Lemon or lime juice, as needed

Ice cubes, as needed

In a gallon glass jar or pitcher, combine the sliced oranges and lime with ¼ cup (50 g) of the sugar. Add the liqueur. Stir until the mixture is syrupy and most of the sugar is dissolved. Chill for at least 2 hours, and up to overnight.

Add the peaches, cherries, herbs, and wine. Stir gently, and chill for 2 hours, or up to 8.

Before serving, taste the sangria and add up to ¼ cup (50 g) more sugar, a tablespoon at a time (orange liqueur varies from brand to brand; some are sweeter than others). If it's too sweet, squirt in some lemon or lime juice.

Remove the herb sprigs if you like, but I think they look pretty in the pitcher.

To serve, add a little of the boozy fruit to each glass. Add ice cubes if it's an especially hot day, or if you plan on drinking a lot. Then pour in your sangria. Garnish with a fresh herb leaf or sprig if you feel like it.

NOTE: Anything left in the pitcher after a get-together will keep well for about 2 days, but I discard the herbs after about 8 hours to keep the works from tasting too grassy.

ALSO TRY WITH: Mangoes, nectarines, apricots, plums, raspberries, blackberries, or blueberries.

PEARS

Pyrus spp.
Rosaceae family
Throughout the US and Canada in zones 5–8

WE LIVED IN A CRUDDY HOUSE in Portland, Oregon, when our daughter was still quite young. When we moved in, our landlord pointed to a lopsided, three-story tree in the back corner of the yard. "That should be a pear tree," he said, perhaps meaning, "That once was a pear tree." A rusty chain choked the trunk about 12 feet (3.7 m) up, and a deflated Mylar balloon ensnared in the tree's topmost branches crinkled in the breeze occasionally.

That first year, three bricklike, lumpy fruits manifested on the tree in the summer and remained resolutely hard. I assumed, just as our landlord did, that the pear tree had given up the ghost. It bloomed dutifully in the spring and shrugged indifferently when the time came for producing anything edible.

Then something happened the following summer: pears. Dozens and dozens of them appeared, from the tree's lowest branches all the way up to the top ones, perhaps to keep the balloon company. Few of the pears grew within reach, so I had to wait until wind or gravity brought them thudding to the ground. I lightly pressed a speckled, yellowish fruit with my thumb and felt the flesh under the thick skin indent just slightly. After a tenuous bite, I ate it all. Delicious.

Faced with hundreds of pears, most of them bruised beyond use by their harrowing fall, I was still able to collect enough from the area underneath the tree to fill up our Radio Flyer wagon. Every day Frances and I would wander out to the tree and begin our process of gathering. With a toddler's love of sorting objects, Frances wholeheartedly applied herself to padding over to the compost bins trip after trip, a pear in each fist, and tossing the damaged fruit in with great finality.

Raccoons nightly feasted on the fallen pears, willy-nilly devouring some sections of the fruit while leaving other parts untouched. Our yard started to smell like a distillery about to be shut down by the health department, and swarming colonies of fruit flies flourished in our rapidly filling compost bins. This is when I realized a person truly does not own a fruit tree. The fruit tree owns *you*.

As the daylight shortened, the daily chore of collecting pears took less and less time. On the final day we found just three, all of them still intact. Frances and I took turns biting from a particularly tasty one, the juice running down my arm as I held the pear out for her to sink her sharp little teeth into. The neglected tree in our unsightly yard was capable of grace and bounty.

Pears are all about good timing. Even though we gladly feasted on many pears fresh from the tree that year, they couldn't have been ideally ripe; pears do not finish ripening on the tree. The pears you bring home are an investment in future fruit snacking.

Pear lovers will accost you with all kinds of pear dos and don'ts: Don't refrigerate them until they are fully ripe; don't pick them before they are ready; don't neglect to pick them when they are ready; don't look at them the wrong way. If you have access to a pear tree, in a

good year, you have plenty of wiggle room, so don't freak out. Utilizing each and every pear at its peak is a full-time job. Accept the collateral damages.

Leave pears on the tree for too long and they will ripen unevenly, until they are rotting from the inside out (I can vouch for this from my experience with my unreachable backyard pears). They also tend to develop a gritty texture. In fact, pears are best when picked just at the first hint of ripeness.

And then, during this off-tree ripening, they have a scandalously narrow window during which they will be perfectly ripe. Pears wait for no one. There's a saying in Spain, *Es la pera*, meant to describe something fantastic. Because if you sink your teeth into a pear during that slender window of perfect ripeness, you will not forget it. It becomes the yardstick to judge all pears that follow.

Today pear production is a very Pacific Northwest thing. Pears are Oregon's number one fruit tree crop. Washington grows more fresh pears for market than any other state. But you can find pear trees growing wherever pear lovers with land can manage to plant them. You won't find any growing wild in North America, unless you count specimens like the one in my former backyard.

The pear is native to Europe and West Asia, but European colonists planted it early on, and it served a great need: Producing beer in the hotter climates and with unfamiliar yeasts was more challenging than the Puritans had assumed it would be. Apple and pear cider proved easier to make and rapidly became the thirst-slaking drink of

choice (well water, remember, was not always safe or palatable, and a low-alcohol fermented beverage was usually healthier).

The Endicott Pear Tree is believed to be the oldest living cultivated fruit tree in North America. Planted by Puritan settler John Endecott (yes, with an *e*) sometime between 1630 and 1649, it grows still in Danvers, Massachusetts, and continues to bear fruit. Now a historic site, the tree is safeguarded by a fence. I wonder if the fruit is any good, and if anyone eats it. The view opposite the tree is a parking lot and a building of medical offices. Some might say it's ironic, but it seems completely natural to me. We came here to build a new world, and so we did.

Ambitious growers of pears might want to consider slipping an empty bottle over a budding branch, so the pear can grow inside as a miraculous garnish, in the style of Poire Williams, the famous pear brandy. Incidentally, what we know in America as Bartlett pears were discovered in 1770 in England and named Williams Bon Chrétien; they are today still called Williams pears in the U.K.

Just as there are cider apples and eating apples, there are perry pears and eating pears. Perry pears are smaller and harder than eating pears, and they're often bitter and packed with astringent tannins. It's unlikely you'll stumble across a pear tree offering perry pears, since nowadays we seem to prefer beverages like Mountain Dew and Vita-minwater. But if you do, you'll know at first bite. They are meant for the noble cause of cider making and not much else.

Harvesting and Storage

When a pear on the tree is no longer green-green and the flesh of its neck gives way to the pressure of your thumb just a teeny bit, it's ready to harvest. The stem should part from the branch with a gentle upward twist; if you have to yank, the fruit is not yet ready. Eschew hard pears. They were picked too soon. Hard pears are different from firm pears, which happen to be firm when ripe.

Pears ripen faster if it's warmer, so for longer storage, ripen pears in a cool, dark place; a warmer spot, like a kitchen counter, will speed up the process. It's okay to refrigerate *fully ripened* pears if you want them to last a few days longer. Pears take well to canning. Freezing, not so much. If you want to dehydrate pear slices, it's best to dip them in a solution of ascorbic acid first.

Culinary Possibilities

A sliced fresh pear likes to be gobbled up completely moments after the knife slides through it. Cut fresh pears will oxidize, so they are not great in fruit salads and such. Given that, I like individual treatments on a whim: sliced pear shingled over toast smeared with quark (a tangy, fresh specialty cheese) is fantastic. Diced pear on hot cereal or cold muesli can make a morning.

Once cooked, their color is locked in, and you have extended their shelf life tenfold. When you come into possession of a bunch of pears that are on the declining side of ripeness, you're best off making them into a puree: pear sauce. From there you can proceed to the classic fallbacks of "too much fruit": Dehydrate it for pear leather, sweeten and freeze it to make pear sorbet, or—my favorite—cook it down to make pear butter, which will easily put apple butter to shame.

Firm-but-ripe pears like to be poached in wine of any stripe: red or white, sweet or dry. They're quite nice this way served over a rich and creamy rice pudding (a classic French dessert called pear condé).

Pears rarely meet a cheese they don't like. Slice them and serve them with a classic cheese fondue for dipping, to offset all that bread. Pungent cheeses in particular are exquisite with pears (Camembert and pears is a classic combo).

ASIAN PEAR

Pyrus pyrifolia
Throughout the US in zones 5–9

A giant Asian pear tree grew—and still grows—at the ramshackle farmhouse where my aunt and uncle lived in the late 1970s, during a brief stint as back-to-the-landers. I have no idea how a mature, fairly exotic fruit tree wound up on century-old farmland in a rural Ohio county, but such is the mystery of inherited fruit trees. Aunt Linda told me about how determined she was to make use of all the Asian pears the tree produced. She made them into decent but not amazing pear sauce and fed it to my then-infant cousin day after day.

To keep up with a productive Asian pear tree, pear sauce is a necessity, but I feel they are best raw. These babies are all about crunch. Unlike the more familiar European pear, Asian pears ripen on the tree and retain a firm, crisp texture. They are round and look somewhat like apples, but there's no apple to them. They have the texture of a jicama and the flavor of a pear, though often more muted. You want to refrigerate them to keep them fresh, but also for the boon of refreshment: a chilled Asian pear is immensely invigorating.

In the South you'll run into what are called sand pears, so named because of their sandy texture. A sand pear is not a species, but more of a collective regional term; some sand pears are hybrids of

P. communis and *P. pyrifolia*, but they are generally more like Asian pears in shape and texture. Because they grow better in hot, humid climates than European pears, they became popular ornamental trees after the Civil War, though their firm texture perplexed those accustomed to the more yielding flesh of *P. communis*. This probably accounts for some unflattering descriptions of sand pears as inedible, or not worth the bother.

Harvesting and Storage

Asian pears ripen in the late summer and early fall. The skin of Asian pears is either russet (a mottled, matte golden brown with slightly raised spots) or smooth and pale yellow with a hint of green. Pick them ripe and sweet, which you can determine by some trial-and-error sampling from time to time. Some may fall off the tree when ripe. Asian pears bruise easily, which is why you see them tucked inside those foam nettings at the grocery store, so handle them with care. Store them in a cool dark place, but ideally refrigerate them; they'll last a week or two that way.

There's no decent consensus about when to pick sand pears. One school says to pick those before truly ripe, to keep the sandy texture at bay. Another says they ripen just fine on the tree. I guess it depends on the variety, as well as the personality of the particular tree itself.

Culinary Possibilities

The charm of Asian pears is their texture. The same high water content that creates the crispness makes them ill suited for some baking applications, though a number of cooks have reported making crisps and such with success. I like them raw in salads and relishes. Substitute them in Latin American recipes calling for jicama. You can even cut them into matchsticks and roll them up in wraps and salad rolls, or grate them and add them to slaws. Their skin is generally thicker than European pears, but it is edible. Peel based on personal preference.

If you're drowning in Asian pears and you don't have a baby to feed pear sauce to, try juicing them into fresh cider, or fermenting that juice into vinegar. Cut them into thin slices and dehydrate them, or go the opposite route and can them (I recommend peeling first). If their favor is fantastic but their texture is too gritty, try making a shrub. And when all else fails, you can always give them away and hope the recipients return the favor by sharing whatever concoction they dream up.

Arugula, Pear, and Almond Pizza

Serves 2–3 as a main, 6 as a starter

You can call this pizza, but it's more like cheesy flatbread with salad on top. The nuts are very important, adding crunch and substance. Smoked almonds are my favorite, but toasted hazelnuts are good, too.

Finely ground cornmeal, for sprinkling

2 pounds (910 g) of your favorite fully risen pizza dough, divided into 2 balls

¼ cup (60 ml) extra-virgin olive oil, divided

1 cup (115 g) coarsely grated Manchego, Gruyère, or Asiago cheese

8 loosely packed cups (140 g) baby arugula or mixed hearty salad greens

2 ripe but firm pears, cored and thinly sliced

½ lemon

Salt and freshly ground black pepper

½ cup (60 g) roughly chopped smoked almonds

Preheat the oven to 500°F (260°C), or as hot as you can get it. Position the racks in the lower and upper thirds of the oven.

Line two baking sheets with parchment paper, and sprinkle each with a little cornmeal. Take the dough and make each round into the pizza shape you like best (I like "rustic blob" shape). Smear about 1 tablespoon of the oil around the outer parts of each dough blob. Press divots into the dough with your fingertips so it won't rise into a big dome when you bake it. Sprinkle the grated cheese over each round of dough, and bake until the cheese bubbles and starts to brown and the crust is browned and its edges have a little char. You want the pizza to be done—it's not baked any more after this. The baking time will depend on the vibe of your oven, but it will probably be 10 to 15 minutes. Swap the baking sheets from top to bottom and front to back midway through baking.

Right before the pizza is due to come out of the oven, put the greens and pears in a large bowl. Drizzle with the remaining 2 tablespoons of olive oil, then squeeze the lemon over it. Season liberally with salt and pepper, and toss. Top the hot pizza with the salad, sprinkle with the chopped nuts, then cut into pieces and dig in.

ALSO TRY WITH: Halved or quartered figs, slices of Asian pear, or pitted and halved sweet cherries.

Asian Pear and Roasted Poblano Salsa

Makes about 2 cups (480 ml)

When summer is on the wane, tomatoes peter out and Asian pears appear. Just go with the flow and put them to work in a crunchy and refreshing salsa. The sweetness of the fruit, the charred bite of the peppers, and the creamy finish of the crumbly mild cheese come together for an enlivening change of pace from the typical salsa. It's worthy of scooping up on those honking big tortilla chips shaped like little baskets.

1 large poblano pepper
3 small Asian pears, cored and cut in
    ½-inch (1.3 cm) dice
½ cup (30 g) chopped fresh cilantro
Juice of 1 lime
½ cup (70 g) crumbled queso fresco
Salt

Heat a broiler or grill to high. Char the pepper, turning it occasionally, until it's black and blistered on nearly every surface. Once it's thoroughly charred, pop it in a paper bag, fold the top over a few times, and allow it to steam for 5 minutes. Remove and discard the charred skin (charred poblano skin can be delicate and difficult to remove neatly, so sometimes I just rinse it off under running water, though this will diminish the pepper's flavor). Then remove and discard the stem, ribs, and seeds.

Cut the pepper in a fine dice, and put it in a medium bowl. Add the pears, cilantro, lime juice, and enough queso fresco to make it interesting. Season to taste with salt.

Serve with corn chips or on top of grilled chicken, pork, or seafood.

NOTE: Usually poblano peppers are on the mild side, but I've had homegrown ones that were blazing hot. Taste a little of your roasted pepper before deciding how much you want to add.

ALSO TRY WITH: Ripe but firm pears, tart and firm apples.

Pear Butter

Makes 5–8 half-pint jars (1.2–2 L)

Fruit butter is a process more than a recipe; it's quite simple, but has a number of distinct phases that are like a Choose Your Own Adventure book (or, if you are an engineer, a flowchart). You begin with a simple puree of cooked fruit, then you sweeten it, and then you cook it down until it's sturdy but still spreadable. Pear butter is my favorite fruit butter. It's suave and silky, and I use a lighter hand with the spices so the pear flavor is nice and strong. A little vanilla bean gets everything to play together nicely and smooths out the sharpness of the ginger.

Your yield will depend on a lot of factors, such as how much sugar you add. I cook my fruit butters down quite a bit to concentrate their natural sweetness before I ever add more than 1 cup (200 g) of sugar. Thus my yield is smaller than a fruit butter with a 1:1 ratio of sugar to fruit puree.

Low-sugar fruit butters are not quite as age-worthy. Give jars as gifts and you'll be popular in your neighborhood forever.

6 pounds (2.7 kg) ripe pears
½ cup (120 ml) water
1 (2 inch/5 cm) piece gingerroot, thinly sliced crosswise
1 (3 inch/7.5 cm) strip lime zest
½ cinnamon stick, preferably Mexican (which is milder)
1 whole star anise
4 cardamom pods
¼ vanilla bean, split in half lengthwise
½ cup (120 ml) lime juice
1–2 cups (200–400 g) granulated sugar

Unless your pears have very leathery skins (or, like the fallen ones from my old yard, are embedded with chunks of dirt), you can just leave them on—no need to peel them. Cut the pears from their cores, and chop them into large chunks. Drop them into a heavy-bottomed stockpot or Dutch oven, and heat over medium-low heat until the pears release their liquid. If necessary, add up to ½ cup water to get them going. Raise the heat to high, cover, and simmer until the flesh and skins are very soft, about an hour.

Meanwhile, gather the ginger, lime zest, cinnamon, anise, cardamom, and vanilla in a square of cheesecloth and tie with cooking twine. Set aside.

Puree the cooked pears with an immersion blender, or in batches in a blender, food processor, or food mill. Return the puree to the pot, and add the lime juice and 1 cup of the sugar. Cook gently for 1 hour, stirring frequently. Remove and discard the cheesecloth bundle.

Continue cooking, stirring frequently, until the color is mahogany and the consistency of the pear butter is thick

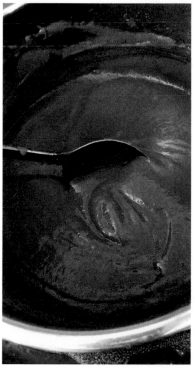

and glossy. Be careful as it simmers in its final stages. It can sputter and shoot lavalike blobs up that are yearning to burn your forearms (or face—really, this happened to me once).

When the pear butter is ready, you can drag a wooden spoon through it and see a distinct trail with ridges on either side. This can take up to another hour or so of cooking. Carefully taste the butter as it cooks down and decide if you'd like to add more sugar; if so, do it in increments of ¼ cup (50 g).

At this final stage, you can puree the pear butter once more for an especially smooth texture.

Pack the pear butter into sterilized half-pint jars, leaving ¼ (6 mm) inch headspace. Process in a water bath canner for 20 minutes. Label and date the jars. Use within a year.

NOTE: You can shave off a little cooking time by initially cooking the raw fruit in a pressure cooker. Put the pears and ½ cup (120 ml) water in the pressure cooker. Lock on the lid, bring to *high* pressure, and reduce the heat to maintain pressure. Cook for 12 minutes, letting the pressure come down naturally. Carefully remove the lid; the pears should be quite soft. Spoon off and reserve any excess cooking liquid, then puree, adding back some of the reserved liquid if necessary to make a smooth puree. You can't cook the pear butter under pressure any more after this first step.

A slow cooker is my favorite method now, because it's so hands-off. Cook the fruit as directed in a slow cooker on *high* (the time it takes will depend on your pears and the heat of your cooker); puree. Then cook for hours and hours over medium heat on the slow cooker. To let steam escape, lay two long wooden spoons across the top of the cooker and set the lid ajar on it. Give it a stir every few hours. How long it will take to get thick and set will depend on how much moisture is in your pear sauce. I cook my fruit butters in the slow cooker anywhere between 24 and 48 hours. This is more than double the time that many slow cooker fruit butter recipes call for, but I want my fruit butters to taste like highly concentrated fruit, not sugar and spice. Remove the cheesecloth bundle after 4 to 8 hours, if you don't want the spices to be too strong.

ALSO TRY WITH: Apricots, peaches, plums, apples. Peel apples, and any other fruit that has thick skins.

Pear Pie

Makes one 9-inch (23 cm) pie, to serve 8–12

Pear pie beats apple pie any day. This has the same lighter spicing as my pear butter, plus a rye pastry to complement the fruit. I like a full double crust on my pear pie, with a sprinkling of coarse sugar to add some sparkle and crunch.

1½–2 recipes Rye Pastry (page 356)

½ cup (110 g) brown sugar (I like dark brown sugar, but take your pick)

Zest of 1 lime or lemon

½ teaspoon Ceylon cinnamon

¼ teaspoon mace or nutmeg

¼ cup (35 g) unbleached all-purpose flour

About 8 large pears, peeled, cored, and cut into ¾-inch (1.9 cm) chunks (8 cups prepped)

1 teaspoon freshly squeezed lemon or lime juice

2 teaspoons vanilla extract

1 egg yolk

1 tablespoon heavy cream

2 teaspoons turbinado sugar, for sprinkling

Preheat the oven to 425°F (220°C). Position the racks in the upper and lower thirds of the oven.

Roll out one disk of Rye Pastry and line a deep 9-inch (23 cm) pie dish with it, letting the pastry overhang by an inch or so. Roll out the remaining pastry disk, and trim to an 8½-inch (22 cm) circle. Refrigerate.

Meanwhile, combine the brown sugar, zest, cinnamon, mace, and flour in a large bowl, and work together with your fingertips until only tiny lumps of sugar remain. Add the chopped peeled pears, lemon or lime juice, and vanilla. Fold to combine.

In a small bowl, whisk the egg yolk and cream.

Scrape the filling into the lined pie dish. Top with the chilled pastry circle. Fold up the overhanging edge, and crimp or pleat as desired. Using a pastry brush, brush the yolk-cream wash over the top of the pie. Cut a few steam vents near the center with a sharp knife, and sprinkle with the turbinado sugar.

Line a rimmed round pizza pan or rimmed baking sheet with foil to catch any of the juices that may bubble over as the pie bakes. Set the pie on the lined pan, and bake on the lower rack for 30 minutes. Reduce the temperature to 375°F (190°C), and move the pie to the upper rack. Bake until the filling bubbles vigorously, 45 to 90 minutes. You may need to tent the pie loosely with foil if the top browns too quickly. Generally speaking, it's better to bake the pie a little longer than to underbake it.

Set the pie on a wire rack, and let cool until warm, at least, before slicing and serving. If you cut into the pie a few hours after it comes out of the oven, the filling will gush all over and the slices will be kinda messy, but they will taste divine—especially with vanilla ice cream.

Patchwork Pies

Any good recipe for pie pastry will leave you with trimmings, because it's really hard to roll out a perfect circle and position it in a pie plate just so. Save those trimmings and freeze them. Once my freezer starts to get a little crazy with parcels of dough wrapped in unmarked wads of plastic wrap, I thaw them and roll them out to make a patchwork pie crust—even if the doughs are not from the same recipe. Patchwork pies give you an impromptu pie with barely any work. If you ever have a fruit windfall, they let you get right down to business and stick a pie or tart in the oven, stat.

Pear Vichyssoise

Serves 8–12 as a first course

Cold potato-leek soup? Boring. Throw a few pears in there to liven it up a bit. This is a good way to put soft or nearly over-the-hill pears to use. Serve it at the start of a swanky dinner party, or be déclassé and sip it from a mug as a snack. I won't tell.

Make this a day before you plan to serve it, so it will be nice and chilled. The pear asserts itself more this way, but make no mistake, this isn't a fruit-forward soup. It's subtle and suave and understated. Traditionally white soups call for white pepper, but black pepper is so much fruitier; I think it's a better flavor match.

2 large leeks, white and light green parts only, well washed and thinly sliced

1 tablespoon butter

2 large (1½ pounds/680 g) russet potatoes, peeled and cut into 1-inch (2.5 cm) chunks

4 cups (960 ml) rich chicken stock (or, for more of a fruit vibe, water)

Salt to taste

1½ pounds (680 g) ripe pears, peeled, cored, and cut into chunks

½ cup (120 ml) crème fraîche

1 teaspoon honey

1 teaspoon apple cider vinegar (hell, use pear vinegar if you have it)

A few grinds of black pepper

Chives, for garnish

Sweat the leeks in the melted butter. Add the potatoes and chicken stock, bring to a boil, reduce the heat to a simmer, and cook until the potatoes are tender, 10 to 20 minutes. Season with salt, and set aside to cool for an hour or so.

Add the pears and crème fraîche, and puree. Thin the soup with water, if needed, so it's creamy and smooth, but not thick and gloppy. Add the honey and vinegar, and then season to taste with salt and pepper.

Chill at least overnight. Taste again before serving (cold foods often need another seasoning bump), and serve garnished with snipped chives (chervil sprigs are nice, too).

PERSIMMONS

Diospyros spp.
Ebenaceae family

A LITTLE ANCIENT GREEK: *dios* means "god"; *pyros*, "grain" or "food." So there's your food of the gods, folks, possibly growing outside your window. You'll find three kinds of persimmons in North America, and though they're all quite different from one another, none of them are the kind of fruit most people pick up and eat a big chunk out of—gods be damned! You have to be intentional with persimmons, engage with them. This probably accounts for why persimmons are not a default member of our nation's oranges-apples-bananas fruit lexicon.

Persimmon trees offer ease of cultivation, culinary possibility, and ornamental beauty—precisely why they are excellent targets for urban foragers from coast to coast. With a harvest period from the end of September to the beginning of December, persimmons can cover all your fall fruit cravings. And they find their soul mates with cinnamon, ginger, clove, and other spices evocative of autumn. Persimmons, by the way, pack up to twice as much fiber as apples, if fiber is the main reason you eat fruit.

Finding persimmons is easy enough: Look to your neighbor, your cousin, your co-worker, or anyone whose yard boasts a persimmon tree. Even a devout lover of persimmons can find themselves swimming in the things if said tree grows on or near their property.

Persimmons are native to Asia, unless they're native to North America. Persimmons of Asian descent are far more accessible and

way less seedy then their New World cousins. But let's begin with those, our own native persimmons, the most persnickety and under-the-radar ones.

AMERICAN PERSIMMONS

D. virginiana
Eastern and central US

American persimmons, which can be found growing wild and cultivated, produce fruit that's the size of bloated acorns. A midwestern delight, they don't grow west of Kansas. They are sweet, sticky, and full of trickiness: They ripen only on the tree and must be collected exactly at the right moment, after they have dropped from its branches but before they have been flattened all over the sidewalk by the soles of clueless pedestrians. That's usually how I identify American persimmon trees: not by peering at up leaves and branches, but by looking down and noticing sidewalk slick and scabby with persimmon smears in October or November. Then I begin staking that tree out.

There's an old saying about American persimmons: Don't eat them until after the first frost. A hard frost can kick-start the bletting process

necessary for the fruit to be palatable, but if you see shriveled-up persimmons that have dropped from the tree, pick one up and taste it. That's the true test.

Harvesting and Storage

Old-timers used to lay a sheet under a persimmon tree at night to collect the fruits as they dropped. In lieu of this, just go back to the same tree over and over during the late autumn and gather the spoils of the day before they begin to spoil. That drop, necessary though it is, can do damage, so refrigerate them as soon as possible, and extract the pulp within 24 hours.

The squishy little buggers house giant seeds, so you get a lower yield on pulp. Accumulating a cup of it is messy and time consuming and takes a good dozen or so fruits. For a small harvest, just obliterate them with your bare hands, and you'll find flesh with a consistency somewhere between marmalade and baby food. Leave the delicate skins on, if you like. I enjoy the hint of texture they lend to cookies and quick breads.

For pulping pounds of American persimmons, your food mill is faster, but count on having to mash those buggers for a good long time to get anywhere.

Culinary Possibilities

American persimmons have dense flesh, with a bit of grit to their texture. Their sweetness is almost cloying, but it is offset with a streak of wild, woodsy smoke. To my mind they are best when baked in dessert preparations. They don't oxidize in their raw form, but when baked into recipes with baking soda they'll turn the whole thing the brown-black hue of molasses over time. Just roll with it.

HACHIYA PERSIMMONS

D. kaki 'Hachiya'
Western US in zones 7–10

If you have licked felt lately, you can identify with the chalky-cheek sensation of biting into an unripe Hachiya persimmon. They're packed with astringent tannins. Get Hachiyas while they are still firm, and they'll grace your counters as they do their thing and ripen into non-astringent, honeyed sweetness. Big, tapered Hachiyas are soft when ready to use, just like American persimmons. Score their skin, and inside you'll find a gelatinous pulp the color of salmon roe.

In 1870 the USDA imported grafted trees and distributed them to California, among other states. Between California's persimmon-

friendly climate and its population of Asian immigrants, the fruit took hold. Those outside of the Golden State may associate it with crates of oranges, but true Californians know the annual autumn showering of persimmons is a far more accurate embodiment of what it means to live there. More temperate regions are lousy with persimmon trees, and often they are Hachiyas. Their annual abundance can provoke the same kind of dread that monster summer garden zucchini do. Anyone with a persimmon tree knows those bushels of fruit can be hard to find good homes for, even though on paper Hachiyas are lovable and easygoing. In reality, they tax our culinary creativity come round three or four. *More* persimmons? Again?

Hoshigaki, Japanese dried Hachiya persimmons, are an increasingly rare delicacy that command up to $35 per pound. A faint, powdery bloom of naturally occurring fructose coats their exteriors, and they are eaten as is for a chewy, flavor-packed treat. Unlike most Hachiya preparations, hoshigaki use still-firm persimmons. As they dry over several weeks, you must massage their flesh every day to break down the cell walls of the fruit. It's a labor of love, but if you face a bushel of homegrown Hachiyas that you can't bear to waste, it's a lovely end for them to meet, given you have success.

Harvesting and Storage

Big ol' knobby Hachiyas are easy to gather. They'll still be firm when you do, which takes the pressure off. To speed up ripening, put Hachiyas in a paper bag with an apple; the ethylene given off by the apple will soften them up. The easiest way to preserve Hachiya persimmons is to freeze the pulp in zip-top bags. You can even freeze an entire ripe fruit, let it thaw a bit, and enjoy the pulp with a grapefruit spoon as a whole-foods take on sorbet.

Culinary Possibilities

You can use Hachiya pulp in any baking recipe as you would mushed-up ripe.

An alternative: fruit leather. Fire up your dehydrators, kiddos. Or whir some pulp in a dairy-licious smoothie, with or without other fall fruits. Generally, you can use Hachiya pulp in recipes calling for American persimmon pulp, and vice versa, but their sugar content and moisture content aren't identical, so you may have to make some adjustments on the fly.

FUYU PERSIMMONS

D. kaki 'Fuyu'
Western US in zones 8–11

Squat, round, and glossy skinned, Fuyus are the most photogenic and user-friendly of persimmons (a friend referred to them as "the kind people actually like"). Fuyus present no fussy "eat only when totally squishy" bletting issues—you can enjoy them while they are firm but ripe; most often they are served sliced. If you've ever spotted persimmons at a grocery store, quite likely they were Fuyus, as they hold up to shipping best.

Fuyus don't need winter frost, love hot summers, and bloom later in the season. Perfect for commercial crops! And they don't need a lot of space to grow and don't require cross-pollination, so a single tree will bear fruit. Perfect for planting at home! Add to this that persimmons are fairly drought resistant, and their affability in landscaping emerges.

Culinary Possibilities

Technically, you could eat a Fuyu like an apple, but most folks cut them into chunks. Wrap a slice with prosciutto for a classy pre-Thanksgiving nibble. Fuyus love salads. Combine them with bitter greens, toasted nuts, and dried fruits or pomegranate seeds in a simple vinaigrette for a harvest-themed salad.

Giant Fall Salad with Fuyu Persimmons

Serves 4–8

Slice-and-eat Fuyu persimmons are a colorful addition to dressy salads. When I see big salads on restaurant menus, they sound good on paper but usually disappoint me by not having enough exciting chunky and crunchy stuff. This salad has lots!

For the vinaigrette
2½ tablespoons red wine vinegar
1 tablespoon balsamic vinegar
2 teaspoons Dijon mustard
½ small shallot, minced
1 teaspoon kosher salt
½ teaspoon freshly ground
    black pepper
¼ cup (60 ml) extra-virgin olive oil

For the salad
8 cups (142 g) mixed salad greens
    (including some bitter varieties like
    arugula and frisée)
8 radishes, very thinly sliced
2–3 Fuyu persimmons, thinly sliced
    crosswise or cut into wedges
2 tablespoons roasted and salted
    sunflower seeds
2 tablespoons roasted and salted
    pumpkin seeds
½ cup (72 g) crumbled fresh chèvre

In a small bowl, combine the vinegars, mustard, shallot, and salt and pepper. Whisk in the oil, and set aside. Dip one of the lettuce leaves in there to take a taste, and adjust the seasonings as necessary.

In a large salad bowl, toss the greens, radishes, and half of the persimmon slices with enough dressing to slick it all up, but not weight it down. Top with the remaining persimmon slices, plus the seeds and the chèvre. Grind some black pepper over the works and serve.

NOTE: You'll probably have some dressing left over. It's versatile, so use it on subsequent salads.

I love adding drained, cooked lentils to this salad. They make it a meal. Nice, firm lentils that keep their shape after cooking (such as lentilles du puy or beluga) are best suited for salad.

ALSO TRY WITH: A handful of pomegranate seeds, or thinly sliced ripe but firm pears or Asian pears.

Persimmon Bread

Makes 1 large loaf

A woman I was talking with at the North Carolina Pawpaw Festival told me she doesn't pulp her American persimmons through a food mill. "I just take out the seeds and leave the skin on," she said. "You get more fruit, and in big, soft chunks." Incidental conversations like that are the whole point of going to events. You meet like-minded people and swap insights and opinions you might not get from a cookbook or an article.

Anyway, I agree with her now. American persimmons have very tender skin, and unless I want a smooth puree, I just tear up my gooey ripe persimmons by hand. They give this sweet and spicy bread a texture that's almost like fruitcake studded with apricot bits. If you prefer a bread with a less rustic texture, you certainly may use sieved persimmon.

1⅓ cups (175 g) unbleached all-purpose flour
½ teaspoon table salt
1 teaspoon baking soda
½ teaspoon cinnamon
¼ teaspoon freshly grated nutmeg
¼ teaspoon allspice
¼ teaspoon cloves
1 cup (240 ml) American persimmon pulp (some soft skin is okay)
1 cup (220 g) brown sugar
2 eggs
¼ cup (60 ml) buttermilk or sour cream
¼ cup (½ stick) unsalted butter, melted and cooled
1 cup (120 g) chopped toasted pecans or walnuts

Preheat the oven to 350°F (175°C). Grease a 9 × 5-inch (23 × 13 cm) or 8½ × 4½-inch (21 × 11 cm) loaf pan.

In a large bowl, whisk together the flour, salt, baking soda, cinnamon, nutmeg, allspice, and cloves. Set aside.

In a medium bowl, whisk together the persimmon, brown sugar, eggs, and buttermilk until well combined. Fold in the flour mixture with a rubber spatula to create a smooth batter. Fold in the melted butter and nuts.

Scrape the batter into the prepared pan, and bake until a toothpick inserted in the center of the bread comes out clean, 50 to 60 minutes. Set the pan on a wire rack to cool for 2 minutes, then remove the bread from the pan, and set on the rack to cool completely.

The bread will keep in a plastic bag for about 3 days, or tightly wrapped and frozen for 3 months.

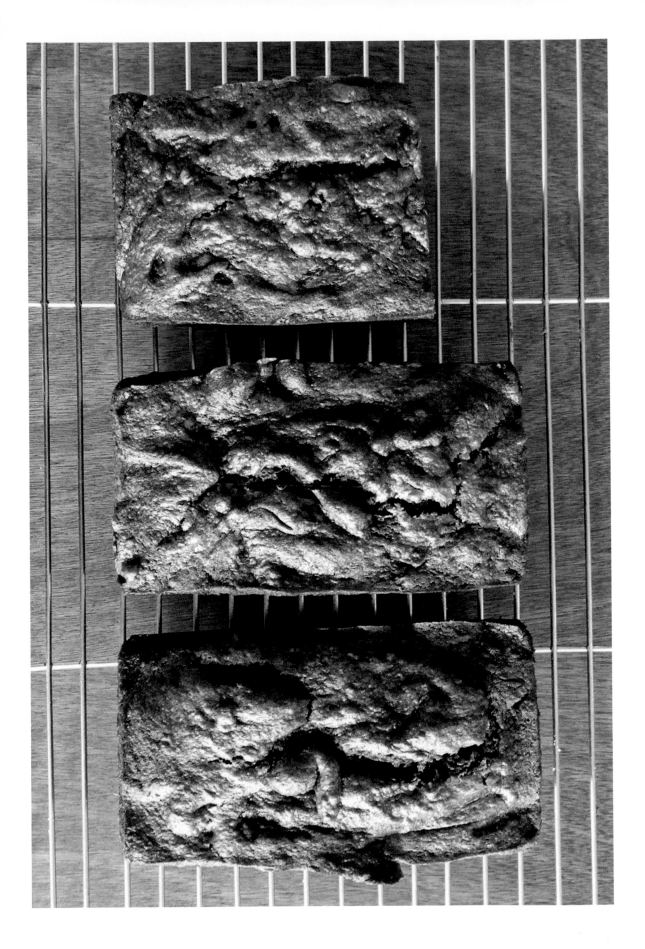

Persimmon Pudding

Serves 12

So much better than pumpkin pie, and easier, too. Break this out for Thanksgiving. You can use Hachiya or American persimmons.

1½ cups (195 g) unbleached all-purpose flour
¼ teaspoon salt
1 teaspoon baking powder
1 teaspoon cinnamon
½ teaspoon ground ginger
¼ teaspoon allspice
2 cups (480 ml) persimmon puree (Hachiya or American)
⅔ cup (130 g) sugar
3 large eggs
1½ cups (360 ml) buttermilk, preferably whole-milk
¼ cup (60 ml) heavy cream
¼ cup (½ stick) unsalted butter, melted and cooled

Preheat the oven to 350°F (175°C), and position a rack in the middle. Grease a 13 × 9-inch (33 × 23 cm) baking dish, preferably glass or ceramic.

In a medium bowl, whisk together the flour, salt, baking powder, cinnamon, ginger, and allspice. Set aside.

In a large bowl, beat the persimmon and sugar. Beat in the eggs, then the buttermilk. Stir in the cream. Fold the flour mixture into the pulp with a rubber spatula until smooth. Stir in the melted butter.

Scrape the batter into the greased dish. Bake until the center is set but still jiggly (like a pumpkin pie), 50 to 60 minutes. The sides will rise up and brown, and the surface of the pudding will be shiny and turn amber. Cool to room temperature, and serve with crème fraîche or whipped cream.

The pudding will keep 2 to 3 days at cool room temperature. I suppose you could refrigerate it, but I like it better when it's not cold.

PLUMS

Prunus spp.
Rosaceae family
Throughout the US and Canada

THE QUESTION WITH PLUMS IS, "Which plum?" They manifest in a rainbow of colors and grab bag of sizes: tapered purple damsons, marblelike cherry plums, glowing green-yellow greengages. There are wild plums for nearly every region imaginable, most of them native and a few of them naturalized. Like children, each has its own personality, and each deserves to be loved without prejudice.

That we even have native plums may come as a surprise. I've always thought of plums as fruits of elsewhere. Plums are adored in eastern Europe, made into brandies and dumplings and sauces. China and Japan are also hotbeds of plum culture. Plum sake is enchanting enough to be dangerous to the weak-willed. In China, delicately pink-white plum blossoms are a flower of winter and symbolize hope, purity, and the promise of a new season.

There is something very mature and sexual about plums when they have swelled into fruit. They are fleshy, sometimes sweet, and never excessively so. There's always a bit of an edge to plums. Not everyone wants to be edgy.

You cannot get any edgier than wild plums. They are a diverse group, but their fruits are usually smaller than their cultivated counterparts. Native Americans used them fresh and dried. Most wild plum trees don't grow terribly tall, between 10 to 20 feet (3 to 6 m) on

average. To rattle off all the species would both overwhelm and dazzle you. A sampling of a few common ones follows.

The American plum, *P. americana*, has the largest range and is found most everywhere but the West Coast. Its branches are thorny, and it often grows in thickets at the fringes of woods, prairies, farms, and fields. Besides thriving in the wild, it is cultivated as an ornamental and can make a good hedgerow. Its fruits are about an inch (2.5 cm) across and red or yellow when ripe.

You'll find native Chickasaw plums, sand plums, or sandhill plums (*P. angustifolia* Marshall) from Pennsylvania to Colorado all the way to the South and Southwest. They grow naturally in sandy soil but aren't too picky to grow elsewhere, and their fruits are similar to *P. americana*.

Beach plums (*P. maritima*) are native to coastal states from Maine to Virginia. True to name, they grow on coastal sand dunes in shrub form up to 7 feet (2.1 m) high, but when growing inland can get up to 18 feet (5.5 m) tall. They are bittersweet and suitable for eating off the branch when quite ripe. They adapted to thrive in their habitat with a waxy coating on their fruit and leaves to protect them from salty air.

For domesticated plums, there are many varieties, too, but it is impossible to glance over the Santa Rosa plum. Plant breeder Luther Burbank had a soft spot for stone fruits, and the Santa Rosa is his most famous hybrid, a produce aisle fixture for decades. Of the over eight hundred plants he introduced in his lifetime, this is the one he named after where he lived and worked. Poet and essayist Amy Glynn grows a tree in her backyard, and writes, "Santa Rosas require little care, bear

heavily (sometimes overwhelmingly), and, maybe because they're naturally full of serotonin or maybe just because they are idyll incarnate when conveyed directly from branch to mouth, they just make you happy." The Santa Rosa plum is the plum we imagine if we are asked to picture a plum. It is large, a glossy dark red on the outside with yellow flesh inside, and sweet enough to entice but tart enough to be interesting. Flashier hybrids have since eclipsed the Santa Rosa as an important truck farm crop, but it continues to be a darling of hobby gardeners and small-scale farmers.

The crux of late summer and early fall is the time for plums. The period when one season overlaps with another is always exciting and confusing, because you are packing up one set of expectations and dusting off another. Where do the plums fit in? Are they a key to enter a realm of sophistication and worldliness? Innocence is overrated. Pick the plums.

The way to use them has to do with how they give up their flesh. Some are impossibly reluctant to separate from their pits; others are so soft when ripe the pit comes squishing right out, leaving a shapeless smear of ripe fruit in your hand. The latter are the sort of plums that grew behind a rental house I stayed in during my salad days—one tree bore yellow ones and another bore red ones, both the size of fat grapes—and at a loss for what to do with them, I took their jamlike raw goop and layered it in coffee cake batter, though messing with those teensy softies was infuriating.

Tklapi, a fruit leather of plums that are cooked and then dried in the sun, is made in eastern Georgia (the Baltic state, not the US one).

It's made from sour tkemali plums and used to flavor stews and soups. Author and food historian Darra Goldstein refers to it as "an incomparable souring agent . . . less astringent than vinegar, more flavorful than tomatoes." The plums also make it into a savory, vaguely ketchuplike sauce served with meats and vegetables. There's a tart version using green plums and a sweeter version made from red plums.

Harvesting and Storage

The plums on a tree will generally ripen all at once. Let them do so, if possible; they will not ripen as well off it. Then be ready for a deluge. Some of them may drop, but the rest should release with a subtle twist. To gauge ripeness, taste, because some plums are green or yellow when ripe. Firmness is also not a terribly useful indicator if the plum tree is unfamiliar.

Very ripe and soft plums spoil quickly. Keep them in a shallow layer and refrigerate; use them within a few days, at the latest. Firm plums can often hang out on the countertop or in a cool, dark spot, where they may continue to grow softer and a tad sweeter.

You can pit and chop plums and freeze them for baking or cooking later. I've never dried plums myself, but since they compose an entire genre of dried fruit, it's worth a shot. Commercial prunes are made from dark and oblong prune plums, which have a lower moisture content. If you dry other plums, they are just dried plums. Both are excellent. Plums with astringent skins are probably best cooked and sieved and made into leather (your own tklapi!) instead. For a tart leather, don't add anything, but if your plums are a little on the puckery side and you want a dessert-style leather, cut it with some applesauce.

Culinary Possibilities

Plum jam and plum butter are preserves to be reckoned with. Their bartering value is high. If you are a preserver, all the clingy and tiny and sour and generally inconvenient plums are your raw material to spin into gold.

Plums do very well on the savory side. Swap firm-ish plums for tomatoes to make fruity relishes, or slip wedges with fresh mozzarella in a riff on caprese salad.

Bakers can load crisps, crumbles, and tarts with sliced plums. Sauté some in butter as an ice cream or pound cake topper, or even backtrack and plop that lovely mess over grilled pork or chicken before you even get to dessert.

Inherited Fruit Trees

The US Census Bureau calculated that the average American will move 11.4 times in a lifetime. I am forty-one years old with a lot of lifetime remaining, and I have already handily beat that statistic.

A well-cared-for apple tree will start bearing fruit two to five years after planting, and the tree itself can thrive for decades. It is quite possible you will at some point live on or near a property with a fruit tree you yourself did not plant. I call these inherited fruit trees. This doesn't really make sense, because if you bought the property you also bought the tree, but it sounds more romantic.

There's something dreamy about having a fruit tree growing on your property—it combines the Steinbeckian idea of living off "the fatta the land" and the epitome of luxury. In 1941 groundbreaking developer Bill Levitt created a planned community, Levittown, on Long Island to house families seeking their own tidy manifestations of the American Dream. His 6 square miles (1,500 ha) and seventeen thousand homes had fruit trees planted in their yards. My uncle, a lifelong Long Islander born in the Baby Boom, told me the first owners of those newly built homes could choose from one of several fruit trees themselves.

Oftentimes those who inherit fruit trees are flummoxed with how to handle their new plant companions. What to do with seven bushels of tiny tart plums? Or quarts and quarts of loquats? Or—this one may be the ultimate test—a truck bed full of Hachiya persimmons?

The smart answer is to give them away. We fruit-loving foragers depend on the kindness and fruit fatigue of disenchanted fruit tree owners. In return, we keep their compost bin from getting overrun. You can be a matchmaker and hook up fruit donors with fruit takers.

Italian Plum Cake

Makes one 10-inch (25 cm) cake, to serve 8–12

Here's a very Euro-style dessert that's a lazy cross between a sponge cake and a torte, yet more rustic than either. The olive oil pushes the fruit front and center. It's adapted from a recipe by Faith Willinger, who advocates tweaking it based on what's in season or handy. This cake is not actually Italian, by the way: I just like the way the name sounds.

¾ cup (100 g) unbleached all-purpose flour

¾ teaspoon baking powder

¼ teaspoon salt

¼ teaspoon finely grated lemon zest

¼ teaspoon cinnamon

1 large egg, at room temperature

½ cup (100 g) granulated sugar

½ cup (120 ml) olive oil

¼ cup (60 ml) milk

½ teaspoon balsamic vinegar

1 pound (455 g) plums, pitted and halved or quartered

2 tablespoons turbinado sugar

Preheat the oven to 350°F (175°C), and position a rack in the center. Line the bottom of a 10-inch (25 cm) springform pan with baking parchment. Grease the sides and bottom well with baking spray or butter. Set aside.

In a medium bowl, whisk together the flour, baking powder, salt, lemon zest, and cinnamon. Set aside.

With an electric mixer, beat the egg and the sugar on high speed until the mixture is creamy, pale yellow, and lighter in volume, about 5 minutes. With the mixer on low, add the olive oil, then the milk and balsamic vinegar. Fold in the flour mixture with a rubber spatula just until it makes a smooth batter.

Scrape the batter into the prepared pan. It will look really skimpy once it's in the pan, but don't worry. Arrange the plums in a single layer across the batter (I like to put them skin side down; you can be as messy or tidy as you like, because the batter will rise up over most of the fruit as it bakes anyway), and sprinkle the cake with the sugar.

Bake for 50 to 55 minutes, until the cake is golden brown on top, a little puffed, and set in the center (a toothpick should come out free of batter but may have a few crumbs clinging to it). Cool on a wire rack for 5 minutes, then remove the sides and cool until just barely warm. You can serve it either that way, or at room temperature. Vanilla ice cream, whipped crème fraîche, or good plain whole-milk yogurt are all very nice accompaniments to this. I also like to pick at it straight from the pan.

NOTE: For an apricot version, omit the cinnamon and balsamic vinegar. Use ¼ cup (60 ml) melted and cooled unsalted butter instead of the olive oil, and add ⅛ teaspoon almond extract. Either halve or quarter the apricots, depending on their size, and arrange them cut side up in the batter. Scatter 2 tablespoons toasted sliced almonds over the top, if you like, and reduce the turbinado sugar to 1 tablespoon.

ALSO TRY WITH: Cherries, strawberries, nectarines, blackberries, or raspberries.

POMEGRANATES

Punica granatum
Lythraceae family
Southern US in zones 8–10

FERTILITY IS FICKLE. It rains down upon women without regard for their needs or dispositions, or it may coldly close itself off to those who yearn for it most deeply. It mocks our schedules and offers no convenient on/off switch. Fertility is a reality, and not a choice.

How alluring, then, to think of the ability to bear a child as a dense, leathery fruit on a branch within reach. You pick it or you don't. There is the thing you ask for, and you get it. If your wish is nothing, the tree grants your wish and allows you to walk away, no questions asked. Free will and divinity are not commonly thought of as bedfellows, but our biological realities give us both, and they do not always get along. Planning a pregnancy (or banking on the absence of one) remains a rocky terrain. A pomegranate reduces the conflict to a something much more gracious.

"The pomegranate seems to have been in existence ever since the earth was created," writes Hildegard Schneider, who was the first head gardener at The Met Cloisters in Harlem. And it is a very, very old fruit, one so striking it worked its way into creation stories and religious texts in nearly every culture exposed to it early on. A pomegranate in a work of art is never just a pomegranate. Its calyx—the persistent and prominent fused sepals on the blossom end—and its five to seven distinct points are said to have inspired the shape of kings' crowns. This alone has influenced everything from chess sets to Burger King.

And we are not yet even inside the fruit, which is shot through with seeds like a pouch of garnets. In the sixth of the Unicorn Tapestries now hanging at The Cloisters, the captive unicorn rests placidly in a pen under the shade of a pomegranate tree, the trunk of which it is chained to. A few of the pomegranates in the tree are split apart and drip their juice onto the coat of the unicorn below. It is an allegory of marriage and fertility, but like all of the Unicorn Tapestries, it is enigmatic. Who is doing the choosing?

As it happens, pomegranates evolved to disperse their seed by cracking open and displaying their treasures to hungry wildlife. They were first domesticated in Persia three to four thousand years ago. A pomegranate's juicy arils are naturally protected by more custom packaging than a flat-screen television, and that feature helped pomegranates journey miles and miles away from their homeland. They were fruits of privilege and luxury for those along trade routes.

Pomegranates in the early twenty-first century went from being a fertility and status symbol to a fashionable health food, thanks in large part to the marketing savvy of billionaire entrepreneur Lynda Resnick, the driving force behind the POM Wonderful brand. She is the reason you can buy refrigerated pomegranate juice year-round in shapely bottles at the grocery store. According to the company's research, before the launch of POM Wonderful in 2002, only 4 percent of the American population had ever tasted a pomegranate. POM Wonderful funded studies about the benefits of the fruit's rich antioxidant content, and Resnick recruited her jet-set social circle to act as influencers. Her 2010 bestselling memoir-*cum*-marketing-handbook is called *Rubies in the Orchard*.

Juice is fine for those who prefer to pound their antioxidants, but picking apart a pomegranate the inefficient way is a medicine of its own. Of all fruits, the sensuality of handling one rent asunder is the most absorbing: the crumbly pith, the tumble of glistening seeds from the thin membranes. I gave one to my young daughter to mess around with and it kept her rapt for ages, though the cleanup took just as long.

A friend visiting from Arizona once showed up with a bag of pomegranates from a tree in her neighborhood, like a Silk Road trader arriving with amazing specimens from a faraway land. I'd never had so many of the fruits at once, and I've still never seen a tree in person—not even from the window of a car hurling down I-5 through California's Central Valley, where most of the US commercial crop grows.

Pomegranates like it hot and dry, but the trees can grow in humid climates with some TLC. Outside of the Central Valley, you are most likely to see pomegranate trees growing as ornamentals in a very scattered Pomegranate Belt from Florida to California. They are deciduous, not terribly tall, capable of thriving for over a hundred years, and often trained to grow as single-trunked trees instead of large shrubs. The fruits ripen in the fall, five to seven months after flowering, and their long keeping quality allows us to enjoy them months after harvest, during those times when the tender fruits of summer are wispy memories.

Even in my lifetime, there was a point when the sight of an actual pomegranate was rare—perhaps limited to reproductions of Renaissance art—but in the ensuing decades every beautifully photographed cooking magazine has featured pomegranate seeds scattered over all kinds of dishes in their Thanksgiving issues. In her 1989 Christmas decorating book, Martha Stewart gilded pomegranates with metallic gold paint. It looks as over the top as it sounds. Of famous women, Martha is doubtlessly one who walks up to a pomegranate tree and sees a choice that is hers to make. Should you opt to pick, I say go for the gold inside.

Harvesting and Storage

Overmature pomegranates burst open and display their guts. This looks cool but isn't the best thing for eating purposes. Uniform color is a sign of ripeness, though the color itself will depend on the cultivar.

More dependable is a drum sound when you thump the fruit. They stop ripening right when you harvest them, which you should do with shears, if you can, to keep the fruit intact.

Pomegranates keep at room temperature for weeks and weeks, which is good news for your seasonal still-life compositions, but don't neglect to break into one every now and then and eat it. Harvested pomegranates improve with some age, but when they begin to look tired and shriveled, sometimes the seeds can be rotten or brown.

The best way to get at lots of seeds without hassle is a matter of debate—or shall we say preference. I've read that you can cut one in half and merely tap on its side with a spoon to knock the seeds loose, but this hasn't worked for me at all. Quartering the fruit and submerging it in a bowl of water is popular, as bits of white pith float to the surface, and burst arils won't squirt you with their juice. This is cumbersome to me. Some people score the pomegranate around its equator and remove the outer pith of one half to reveal the seeds inside, like a brain surgeon does with a skull. There's not a wrong way or a right way, and there's certainly not a fast way. Why must everything be fast and easy? What are we in such a rush for when we are handling sexy and colorful fruit? Handling it is the point. Get in there.

Once you liberate the seeds, refrigerate them in a covered container and use them within a few days; they can get brown and squishy before you know it. For juice, I've read that putting halved fruit in a citrus press works quite well. Plopping cleaned seeds in a blender does, too. Strain out the solids and you're good to go.

Culinary Possibilities

When you are tired of garnishing the most beautiful fall and winter salads with strewn pomegranate seeds or dappling your morning muesli with them, juice the suckers. Refrigerated, the juice should keep for a week.

Cooking pomegranate juice down to pomegranate molasses makes it much more useful for cooking and will set you up for a gratifying tour of Persian cookery. There are a few ways to go about it—adding sugar and lemon juice will give it more dimension—but you can do a straight reduction and see how you like it. A sweeter cousin of pomegranate molasses is grenadine, gateway to two famous mocktails: the Shirley Temple and the Roy Rogers (the latter is really nothing but an upgraded Cherry Coke).

Pomegranate Molasses

Makes ⅓–½ cup (80–120 ml)

The trick to this is getting it puckery-sweet. You'll need to fiddle with the sugar and lemon juice, which is why I add it at the end, once the syrup has reduced. A small batch has a low yield, but it comes together quickly because a smaller volume reduces quickly. If you are making a ton of pomegranate molasses, you might want to do it in two batches.

2 cups (480 ml) pomegranate juice
Up to 2 tablespoons granulated sugar
1 teaspoon lemon juice, optional

Put the pomegranate juice in a wide saucepan or high-sided skillet. Bring it to a boil, and cook until it reduces by at least half. Swirl the pan from time to time to see how it's coming along, and reduce the heat as needed to keep it from scorching. The total amount of cooking time will depend on the surface area of the pan you use, but 15 to 20 minutes is realistic for 2 cups.

Once the juice gets syrupy, taste it. If needed, add sugar or lemon juice in increments. How much you'll use depends on the natural sweetness of the juice you started out with, plus your own personal preference. The lemon juice gives it a brighter taste, I think.

The reduction will get stickier and more syrupy as it cools. Add a little water if it's too stiff. Pour it into a jar or bottle, and refrigerate for a month at least, possibly longer.

NOTE: What do you do with this now? Add it to dressings for grain or lentil salads, use it in glazes for meat, or add it to long-simmered Persian stews. Toss root vegetables with a drizzle before or after roasting.

You can also mix this into drinks—iced tea, club soda. It sounds crazy, but if you are a Coca-Cola drinker, try a drizzle of it in there; it'll taste like Cherry Coke. Grenadine, the pomegranate syrup famous in Shirley Temples, has a much higher ratio of sugar (1 part sugar to 1 part juice).

Roasted Root Vegetables with Pomegranate Drizzle

Serves 4 as a side

All fall and winter long, roasted vegetables are a go-to at our house. I can eat half a giant pan on my own. Finishing them with a modest slick of pomegranate molasses brings out their natural sweetness, gives them a foxy burnished look, and adds just enough sourness to keep them from being dessert-y.

12 ounces (340 g) carrots, peeled and cut into 1-inch (2.5 cm) chunks

12 ounces (340 g) beets, peeled and cut into 1-inch (2.5 cm) chunks

12 ounces (340 g) parsnips or sweet potatoes, peeled and cut into 1-inch (2.5 cm) chunks

1½ tablespoons extra-virgin olive oil

Salt and freshly ground pepper to taste

About 2 teaspoons Pomegranate Molasses (page 299)

Preheat the own to 425°F (220°C). If you have a convection setting, use it.

On a rimmed baking sheet, combine the carrots, beets, and parsnips or sweet potatoes. Drizzle with the olive oil, season with salt and pepper, and toss to combine. Roast for 20 minutes, toss, and continue roasting until the vegetables are nice and brown in spots and cooked all the way through, 10 to 20 minutes longer.

Drizzle with the pomegranate molasses, and toss to coat. How much you'll use is up to you, and also will depend on how sweet or tart your pomegranate molasses is. Serve hot or at room temperature.

NOTE: Crumble feta cheese on these for a simple, satisfying meal.

PRICKLY PEARS

Opuntia spp.
Cactaceae family
Throughout the US and Canada

PRICKLY PEAR CACTI HAVE A WIDE natural range. They're native to the southern United States, from the Pacific to the Atlantic, and they provide a fruit and a vegetable all in one plant. *O. humifusa*, the eastern prickly pear, even grows as far north as Massachusetts. You'll see prickly pears in quantity in hot regions, whether humid or dry, growing in coastal areas, rocky deserts, and planters in shopping mall parking lots. Some species were introduced from Central America as ornamentals. If you live in California or the Southwest, there is no shortage of prickly pear cacti to take in—perhaps in your front yard or just down the sidewalk.

The large, fleshy paddles are modified branches, and their spines are modified leaves. What distinguishes *Opuntia* species from other cacti are barely visible hairlike spines called *glochids*. One experience picking prickly pears with your bare hands is enough to emblazon that on your memory. They embed in your skin like fiberglass slivers. It is very unpleasant, but there are workarounds, because otherwise no one would have lasted long eating prickly pears, and humans have used them as a food source for thousands of years. If you've ever had Mexican-style nopales, you've had the paddles. Once cleared of their spines, they are blanched and then sautéed, served warm or cold. They are somewhat bland and, once prepared, are reminiscent of slimy green beans, though they're better than I'm setting you up for.

Prickly pear cactus plays host to the cochineal beetle, colorant of red items such as lipstick and Campari. The nymphs on the paddles secrete a waxy substance that give the host cacti a frosted, blotchy white look.

Periodically—usually in the fall—oval fruits appear on the paddles. Their color may entice, but keep your hands off! If they can make your hands feel awful, imagine what they'd do inside your body. All this fear-mongering is just to say that following the proper precautions will remove the intimidation, and you'll soon be enjoying juicy, colorful free fruit.

Harvesting and Storage

Ripe prickly pears can be yellow, orange, red, or reddish purple. Since there are so many species growing in so many locations, their appearance when ripe will differ, as will their flavor.

You can wear gloves when harvesting prickly pears, but it's possible to improvise by just using an old towel or a few sheets of newspaper to protect your hand. Or grasp the fruit with a pair of tongs and twist it gently. Drop it into a bag or bucket.

Those glochids need to come off before you use the fruit. Passing them through an open flame is a classic method—spear the prickly pear on a skewer or fork and char all of the glochids, including the ones at the top and bottom of the fruit. Blackened spots mean the glochids are gone.

Forager and author Pascal Baudar developed his own technique. He removes the glochids before picking the fruit by making a bundle of small dry twigs or branches lying around and slapping the cactus fruit with it. Baudar says that by using reasonable force but not rough-housing, 10 seconds of swatting will remove nearly all of the glochids. He also cautions not to stand downwind while doing this, or the glochids may blow onto your clothing. Soak the prickly pears in water for an hour once you get home so any stray glochids will soften enough that they won't pierce your skin.

Once separated from the plant, prickly pears do not last long. Refrigerate them and use them within a week. Prickly pears are very seedy. You can't really chew the seeds, but you can swallow them, spit them out, or strain them out. Native Americans saved the seeds, dried them, and ground them into flour.

Culinary Possibilities

Purees and juices are the starting point for desserts, candies, and slaking drinks. Halve the prickly pears, and scoop out their seedy pulp with a spoon. Quickly blitz in a blender until liquefied, then pour through a sieve; the juice will pass right through. Much easier than mudding about with a food mill!

Use the juice as a base for sorbets, ice pops, or a geléelike candy. Of course there is prickly pear jam and wine, too. The most straightforward use is to just drink it, sweetened or unsweetened, over ice or not.

GINKGO

Ginkgo biloba
Ginkgoaceae family
Throughout the US and Canada in zones 3–8

PU! What's that smell? Old socks or unwashed roller derby gear? Nope, it's just the fruit of the ginkgo tree.

Ginkgo trees are awesome, though certainly not native. They originated in Asia some two hundred million years ago; people call them living fossils because they've changed so little in that impressive time span. Their fan-shaped leaves are lovely, turning a beautiful yellow-gold and carpeting the ground when they drop in the fall. They line avenues and dot parks. And the female trees produce fleshy, pale orange seeds in fleshy coats (technically not fruits) that urban dwellers bitch about every autumn, because they stink. Not as bad as rotten death, but definitely in league with vomit.

There's a theory that ginkgo trees developed this smell to attract now-extinct mammals or dinosaurs to consume the berries and then disperse their seeds. The common dispersal method now is different: Modern humans plant seedlings as landscaping features in public spaces because ginkgos are so dang resilient, withstanding air pollution, soil compaction, salt, and other hazards of city life that tend to reduce the life span of trees.

Some cities ban the planting of female trees to reduce the annual putrid stench, though wily ginkgo trees can change their sex—an adaptation to correct an

imbalanced sex ratio. No wonder ginkgos are still around after dinosaur time!

If you are willing, you can investigate those smelly, berrylike seed cones—hitherto referred to as nuts, even though they are not—because you can eat them. In China, Japan, and other Asian countries, they are added to savory and sweet dishes in small amounts. In an article for *Serious Eats*, Chichi Wang described ginkgo nuts as "the Camembert of nuts," nailing the stinky cheese vibe ginkgo seeds have to offer. In communities with a substantial Asian population, you may spot fellow foragers out collecting the ginkgo harvest (all of the aunties and grannies were out foraging long before it was trendy—ask them for tips!).

Know up front that handling the fleshy pulp (the sarcotesta) outside the nut can irritate your skin, causing minor blisters. Wear gloves when handling them, or collect them with your hands shrouded in plastic bags like us pet owners do when we pick up after our dogs.

Also, don't eat the fleshy part, because it's mildly toxic. After collecting fallen ginkgo nuts, remove the pulp and discard it. The seeds inside are your score. Wash them off thoroughly to remove any clinging pulp. Do this all outside if you're smart so you don't faint from the smell.

Ginkgo nuts are not edible raw. You'll need to boil the pulp-free nuts for 5 to 10 minutes before shelling them and eating the soft meat inside (southerners may have some familiarity with this concept via boiled peanuts). Or you can roast them in a 300°F (150°C) oven for 30 minutes.

One last warning: Don't binge on prepared ginkgo nuts. Ginkgo is powerful stuff with a lot of medicinal properties, and it's not advised to eat more than 10 prepared ginkgo nuts a day. If you are new to ginkgo, collect just a dozen and see if you like them before going bonkers and clearing an entire park of its dropped ginkgo nuts. They are certainly not for everyone—which, for a special sort of person, is the ultimate invitation.

Prickly Pear Agua Fresca

Makes about 2 quarts (2 L)

In Mexican markets, you can find cultivated prickly pears, called tunas. They are larger than foraged ones, but either can be used to make this brightly colored drink. Prickly pears have a viscosity that takes some getting used to, but it makes their juice both refreshing and substantial.

3 pounds (1.4 kg) prickly pears, cleared of glochids

5 cups (1.2 L) water

1 tablespoon granulated sugar

Halve the prickly pears, and scoop their flesh into a blender. Pulse until liquefied (the seeds will stay intact). Strain into a pitcher, add the water and sugar, and stir until the sugar is dissolved. Serve over ice, or cut with fizzy water.

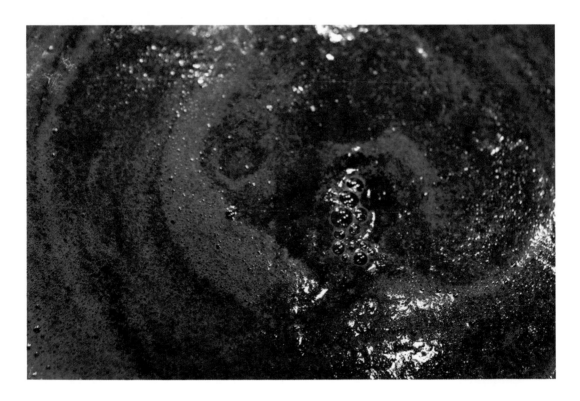

QUINCE

Cydonia oblonga
Rosaceae family
Throughout the US and Canada in zones 5–8

QUINCE IS HARD TO LOVE, and that's because it's not easy to eat. Unlike an apple or pear, you can't simply pluck one from a tree and sink your teeth right in; its flesh is bitter as all get-out, and its skin is flocked and fuzzy. There's peeling involved, plus cooking, plus the addition of sugar. Lots of sugar.

For all the fuss they require, quinces beguile, offering aromatic floral and honey notes. And their resilient off-white flesh transforms once exposed to heat, developing a deep rosy blush and a supple texture. Once you fall for quince, you'll seek them with a passion.

There's an old story in the Bible about a lady named Eve tempting some guy with an apple. Scholars speculate the fated fruit was actually a quince, though I personally doubt it.

Eve: Try this, it's awesome.
Adam: What the hell am I supposed to do with that thing?

Quince is native to western Asia, and it was cultivated earlier than apples in Mesopotamia (now Iraq). It developed to thrive in dry, hot climates and acidic soil, though most of the trees growing in North America were selected to withstand cooler, wetter conditions. It is possible for some varieties of quince to be juicy and palatable right from the branch in subtropical climates, but such examples are needles in haystacks. You must first find a quince tree, period. Quince production in America exists but is uncommon enough that the USDA Agricultural Statistics Service does not track it.

Commercial pectin may be partly to account for quince's rarity in America. Early Americans embraced quince as a useful addition in jams and jellies; it's high in pectin and helps preserves set. In the early twentieth century, it began to fade away as a presence on farms as growing apples and pears became more lucrative. These ready-to-eat fruits, combined with ready-to-use commercial pectin (first patented in America in 1913), were a one-two punch that pushed quince to the fringe.

My dog is the reason I became obsessed with quince. I took him on snaking walks around the blocks near our house in Portland, Oregon, twice a day, and one day I looked at two trees in a small yard we passed not infrequently. Big, knobby fruits like misshapen pears hung from the tree. Quince! It seemed so strange a find in our unremarkable neighborhood full of potholes and dead-end streets. I nabbed three and stuffed them in my jacket pockets. Every day for the next few weeks I'd return and scrump a few more. This wasn't as easy as it sounds; the house was on a corner lot at an intersection with a decent amount of traffic. As I reached for upper branches, keeping my eye out for watchful drivers, Scooter would pace and wind his leash around the tree trunks and whine.

I think I liked the impropriety of it, the small thrill of being bad. Those quince-lifting walks were the highlight of my day, the first thing I thought about when I woke up. Eventually I stockpiled enough to fuel all kinds of cooking experiments: quince jelly, poached quince, roasted quince, quince tatin, and the dense quince paste known as membrillo.

This one-woman festival of quince became an annual rite, until one year I decided I'd had enough unnecessary furtiveness—a few times I skipped my quince stealing because the house's window blinds were open—and I walked up and knocked on the door. "Hi," I said nervously. "I live in the neighborhood. Is it okay if I take some quince from your trees?"

"Oh, so that's what those are," the lady who'd answered my knock said. "Sure, help yourself." And I did. After the adrenaline rush of rule breaking, I found the contentment of human connection, which was better. I offered to bring her a jar of quince jelly, but she declined.

Quince takes well to spices like clove, cinnamon, and nutmeg. Cooked quince is firm but tender and sweet but bracing, qualities that make it excellent to pair with meats, particularly in Middle Eastern cuisines that embrace sweet-sour flavors in rich main dishes.

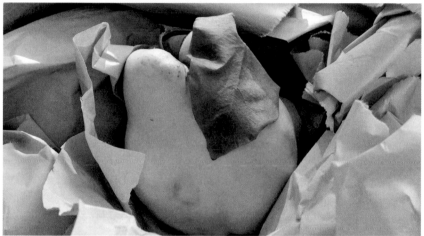

May I be your Eve? We now live states away from those twin quince trees, and I can tempt you with knowledge only. The quince you must seek yourself. Its metamorphosis from the most astringent of Rosaceae to an otherworldly delight will be all the sweeter for it.

Harvesting and Storage

Once it's ready to pick, quince will be a luminous but pale yellow color, and the flocking on its skin won't be so prominent. The fruits become mellower the longer they stay on the tree. Their form may be burly, but they are more easily jostled than you might imagine, so try not to be rough with them as you fill your pockets or bags.

Quince keeps at cool room temperature for weeks. Use it for decoration: I heap them in a large bowl and gaze at it affectionately.

Quince are hauntingly aromatic, too, and if you get close to the bowl you'll be rewarded with their perfume. There's nothing else like it.

Dark brown bruised spots may appear over time. If so, use the marred fruits first; trim off and discard the bruised flesh. Cutting into the tough interior of a raw quince takes muscle. I hope you have a big, sharp knife.

Culinary Possibilities

The fruit's taste is so distinctive it's pointless to even consider pairings like lavender-quince or rosemary-quince; that may be too baroque. (A vanilla bean, however, will soften and lift up those floral qualities. That's the route to go.) As mentioned, quince is packed with pectin, the substance that makes jams and jellies set. Quince jelly is labor-intensive, messy, and worth doing at least once.

The longer it cooks, the rosier a quince's flesh will be. They owe this flattering trait to an unflattering trait: the tannins that make the raw fruits taste puckery. When raw, color-producing pigments are bonded with the tannins in the fruit, the application of heat releases them and destroys some of the tannins, softening the flesh and flavor. Roasted quince gets a nice ruby color, but I find that dry-heat cooking exaggerates its somewhat sandy texture. Poaching, though, renders quince almost custardy.

An entire pie of quince is a bit much even for me, but you can get kinky and mix slices of quince in with an apple or pear pie. Membrillo is fun to make if you are drowning in quince. Unless you own a restaurant or are from Spain, an entire pan of membrillo is a lot of membrillo. It's typically served with Manchego or other semi-hard cheeses. Quince finds its soul mate in dairy products, as it turns out: poached or candied quince with yogurt, quince jelly on toast with lots of butter.

Chunks of quince make the intricately spiced lamb stews of Persia and Morocco even more enticing. Another approach is to invert the combination and stuff halved quince with seasoned ground lamb.

I am always on the lookout for quince trees, yet have not found any nearby despite persistently asking around. I pined for quince so badly I finally bit the bullet and had some shipped to me from an orchard in Oregon. When I opened the box—each quince was cocooned in paper for padding—the smell hit me like a wave, and the glory of all the quince fixings I would make tricked me into thinking I could once again take off to my godchild trees a few short blocks away and snatch some more to go with them.

JAPONICA

Chaenomeles japonica
Rosaceae family
Throughout the US in zones 5–9

This exceptionally pretty shrub is known as flowering quince or ornamental quince, though it could just as easily be called false quince. Like bona fide quince, it's in the rose family, and it's even been classified in the same genus (*Cydonia*) at times. Japonica (or japonica quince, as it's sometimes called) bears an edible fruit that is like a smaller, less fragrant, less compelling version of (for lack of a better term) culinary quince. A japonica shrub, even a big one, will not produce nearly as many fruits as a quince tree. You'd be lucky to get enough to make into anything much.

Native to—you guessed it—Japan, what japonica does have to offer are hundreds of lipstick-pink blossoms in the springtime. I used to have one growing outside my kitchen window, and the quiet joy of seeing it blooming in the cool, wet days of spring easily justified having that shrub in the yard. The fruits, if there are any, will appear in the fall. Like their more flavorful culinary quince cousins, japonica quince will be yellow with a greenish tint when ready to pick. Sometimes the branches are thorny, so take care. Japonicas are not for eating straight from the branch. You must cook them. If you have some knocking around, peel them and cube them and toss them into some wonderful thing you're baking with apples or pears. People do use them to make jelly, and it could quite possibly be amazing, but I'd rather channel my energies into pursuing other fruits. Think of japonica quinces as an unexpected bonus from a plant widely loved for other reasons.

Quince Jelly

Makes 10–14 four-ounce jars (1.2–1.7 L)

Quince lends itself beautifully to jelly making. Since quince has such a distinctive flavor, there's no need (or point) to experiment with fancy-pants flavor combinations. Maybe a vanilla bean, if you're feeling cheeky, but that's it.

Four-ounce jars are the perfect size for quince jelly. Its flavor is too unique to work on PB&J, though it goes well with almond butter. I like quince jelly best on buttered toast or scones.

3½ pounds (1.6 kg) quince
7 cups (1.7 L) water
About 4 cups (800 g) sugar

Rinse off the quince. Quarter them, leaving the cores in and skins on. Put them in a large Dutch oven or heavy-bottomed stockpot, and add enough water to cover the pieces by an inch (2.5 cm).

Bring to a boil, reduce the heat to a simmer, cover, and cook until the quince pieces are fall-apart tender, 45 to 60 minutes. Remove the pot from the heat, and mash the quince with a potato masher until the contents of the pot resemble rough, runny applesauce. (Don't overdo it or you won't get enough juice.)

Put the mixture in a large jelly bag (or a strainer lined with two layers of cheesecloth) and set over a large bowl. Strain for 3 to 4 hours. Measure the liquid; you should have 5 to 6 cups (1.2 to 1.5 L). If you have under 4 cups (960 ml), add a little more water to the mashed quince, and let it strain out again.

Put two small plates, preferably white or light-colored, in the freezer.

Pour the strained liquid into a heavy-bottomed pot. Add a scant cup (200 g) sugar for every cup (240 ml) of quince juice. Bring to a boil, stirring frequently, until the sugar is dissolved. Reduce the heat to a hard simmer. Clip a candy thermometer to the side of the pot to monitor its temperature.

Skim off and discard any scummy foam that may rise to the surface. Continue cooking the liquid until it passes the wrinkle test: A dab set on the frozen plate and allowed to cool for a minute should wrinkle when you nudge it with your fingertip. If it's runny, keep on cooking. As the temperature on the thermometer climbs closer to 220°F (105°C), begin doing the gel test.

Ladle the jelly into sterilized jars, leaving ¼ inch (6 mm) headspace. Screw the lids on fingertip tight, and process in a water bath canner for 10 minutes. Let cool, then label and date.

Poached Quince

Makes 3–4 cups (720–960 ml)

You can roast quince, sauté it, or cook it to death and make membrillo, but all in all I'm a dedicated poacher. It's lazy preserving, more or less, and perhaps the best vehicle for the true flavor of quince. When I get my hands on some, this is the first thing I do. It's not particularly challenging, and the results can top ice cream, yogurt, dense pound cake, or delicate panna cotta.

1½ pounds (680 g) quince
  (3 large quince)
⅓ cup (65 g) granulated sugar
1 (4 inch/10 cm) cinnamon stick
1 (4 inch/10 cm) strip lemon zest
½ vanilla bean
1 small bay leaf
About 2 cups (480 ml) water, or
  light-bodied white or rosé wine
1–3 tablespoons honey

Rinse off the quince, then peel them (a serrated peeler is especially nice for this). Core the quinces, and chop them into ½-inch (1.3 cm) pieces. You should have about 4 cups. (The cut quince may oxidize, but don't worry.)

Place the chopped quince, sugar, cinnamon stick, and lemon zest in a medium saucepan. Split the vanilla bean lengthwise; scrape out the seeds, and add the seeds and bean to the pot. Add the bay leaf, plus just enough water or wine to cover the quince.

Bring to a boil over high heat, then reduce to a gentle simmer. Cover, and poach for 10 to 25 minutes, until a fork easily pierces the quince (the cooking time can vary a lot for quince; start checking after 10 minutes). Remove the quince with a slotted spoon; set aside.

Add 1 tablespoon of the honey to the poaching liquid. Bring to a boil, and reduce until a rosy, light-bodied syrup forms (if the syrup reduces too much, just add a little water). Cool a bit and taste; adjust the sweetness with more honey, if desired. Discard the cinnamon stick and lemon peel, though I do like to leave in the vanilla bean. Pour the syrup over the quince; cool. Store, refrigerated, in a tightly covered jar for up to a month.

NOTE: Pectin-rich poached quince makes for a pillowy-smooth base for a sorbet. Just puree the poached quince with some of its poaching liquid and churn away in your ice cream maker.

RASPBERRIES

Rubus spp.
Rosaceae family
Throughout the US and Canada

A LITTLE SECRET: Blackberries are fine, but raspberries are a million times better.

We lived for a while across the street from an elementary school where the PTO built a garden. It included a few thriving red raspberry bushes, but by the time the berries were ripe, it was high summer and school had been out for weeks. We took our daughter, then still a toddler, to the school playground almost every day, and descending on those berries was irresistible—no one else was picking them! I finally encountered someone filling a small plastic tub and asked her if she knew anything about the berries.

"Sure, I teach here," she replied. "Pick away!" We did. A compulsion overtakes me when I pick berries and I let it carry me away, half ignoring my small daughter as she darted in and out of the brambles like a rabbit. I pretended I was picking black raspberries with my dad at the old farmhouse where we used to camp; I pretended I was picking blackberries with my old boyfriend in a holler in West Virginia after a dip in a swimming hole; I pretended I was picking blueberries on a mountain trail in the Catskills with my husband. I was doing everything I'd ever loved at once, and felt free. When your first child is very young and you are not acclimated to the limitations it places on your comings and goings, sensations of freedom are as good as gold.

We had 2 pints (960 ml) by the time we left. My daughter devoured handfuls of raspberries with dirty hands for dinner. The playground was wild and we were the happy family we wanted to be. Every time I figured we were done, a giant hidden raspberry would catch my eye and I'd be at it again. It's part of our instinct not to give up, even when giving up is what makes the most sense. Maybe that's why brambles have thorns. It's not to keep us away, but to keep us there.

Raspberries are caneberries, along with blackberries, boysenberries, loganberries, and other hybrids and variants. (Why are they called caneberries? They grow on thin woody canes, as opposed to vines.) Look closely at the fruit, and you'll see raspberries have tiny hairs, remains from the pistils of their blossom days. Early European settlers of North America did not adore red raspberries (*R. idaeus*) and regarded them as weeds, but they got with the program soon enough.

Red raspberries are the touchstones, but there are others. Black raspberries (*R. occidentalis*) are sometimes called black caps. Every time I eat a blackberry, I wish it were a black raspberry instead. They are my all-time favorite *Rubus*, with a flavor more complex than reds. They are also somewhat seedier, too. In the midsummer I run into black raspberries here and there on some of my favorite walking routes where brush meets path. There are never enough for more than a passing snack, but I am thankful for that; they are always delicious.

Raspberries have regional bramble cousins to find growing wild. Petite thimbleberries (*R. parviflorus*) are red, with a delicate, velvety

texture. The intensity of their flavor can vary a lot. Of all the *Rubus* these turn to mush the fastest, particularly if you are shoving your fingers in them like they are real thimbles (it's fun, though). Thimbleberry leaves are broad, with soft hairs on the top and bottom, and are velvety like the berries. Salmonberries (*R. spectabilis*) grow in the Pacific Northwest (a berry lover's heaven) and don't look like salmon, per se, but they can be the orange color of salmon roe. Native American tribes would eat the shoots in the spring with salmon.

Wineberries (*R. phoenicolasius*), or Japanese wineberries, are from Asia and have naturalized, becoming invasive in some areas. All the more excuse to pick them! They grow in the eastern United States from Michigan down to Georgia. The developing berries have a thistly, sticky calyx that's equal parts neato and foreboding. The darkest berries are the sweetest.

Harvesting and Storage

It's easy to tell when raspberries are ripe—they taste great, and you will not want to stop popping them directly in your mouth as you pick them. Color isn't an indication, because black raspberries are red before they ripen. They should release with ease from their receptacles, giving them their trademark bowl-like hollow. This depression also makes them more delicate, so don't pile them on top of one another as you pick. If you don't have paper cartons handy, use old metal baking

Boom and Bust Cycles

After a few years of foraging, you'll notice how different harvests can be from one year to another. As I write this, it's a blowout season for peaches in my town. Last year, there were none.

Desirable plants for foraging have a knack for totally missing the happy medium. You might like predictably manageable yields, but a plant does not give a crap what you think. Drowning in

lemons from your two backyard trees? Enjoy it, because I guarantee a year will come when you'll be glad to see a harvest of three lemons, total.

Harvesting foraged fruit is so seasonal, but keeping tabs on the plants throughout the year is something you can benefit from. You'll notice cycles beyond seasons, things that reflect weather patterns and physical changes, too. My favorite patch of black raspberries got mowed down last summer to create an access road for the power company, but guess what? It grew back enough for me to pick a few handfuls this year. I'm curious if it'll get mowed again. As years pass, I feel bonded with the different plants I check in on. I have my own boom and bust cycles, too. I see plants bounce back from lean years with heavily laden branches, and it reminds me to keep plugging away.

pans for collecting the berries, because they are nice and shallow, keeping the top layer of berries from impacting the bottommost ones.

Keep the berries out of direct light, if possible, as you pick, and then get those little guys in the fridge as soon as possible. Freeze them or make them into something within 24 hours of harvesting, unless you want to pitch a moldy pile of mush. I like to set aside a quart or so of fresh berries for nibbling and cereal topping, but even those don't last long.

Raspberries are ill suited for dehydrating, but they can be combined with applesauce for a crowd-pleasing fruit leather.

Culinary Possibilities

There's not much you can't do with raspberries. Eat them or freeze them or put them up. But most of all, find them and pick them and pick them again.

The Best Places for Fruit Can Be the Worst Places for You

I hate giant shopping centers and the vast parking lots that surround them. Who doesn't? People drive around them all crazy because often they're designed with no concern for traffic flow or pedestrian access. In the summer the dark pavement is unrelentingly hot and sticky; in the winter heaps of dirty plowed snow make collars around light posts like pathetic faux glaciers. In all seasons litter collects along curbs and in drains. You have to get through this gauntlet of flotsam before entering the big-box stores themselves, our air-conditioned temples to modern consumerism, wherein clean and colorful merchandise beckons.

I go to these places because I have to, not because I want to, but a few years ago I noticed something that started making these reluctant shopping trips more tolerable. Where there's a giant shopping center, there's disturbed soil . . . and where there's disturbed soil, there are tenacious plants. I make a point of parking on the periphery of the lot to check out what's thriving there. In the summertime particularly, a whole new world emerges: rampant thickets of Japanese knotweed, cruel brambles of blackberries, stunted crab apple trees. I like to think of these terrains as a glimpse of a world without humans. Plants were here before us, and unless we manage to blow up our planet in one clean, careless blast, plants will be here after us, too.

If I find actual fruit in such spots, I barely pick it, and I rarely eat it (who knows what nasty crud might be contaminating the stuff!); it's a tiny foraging trip to satisfy my curiosity, not my appetite.

Fast-food restaurants right at interstate exits are great for such outings, too. Let's say you're traveling and need to eat and you didn't bother to pack real food with you. Once you've made the unglamorous choice among the Subway, Burger King, and Wendy's in the middle of nowhere, park your car and see what's growing a few feet from the dumpster. Sometimes there's a bona fide field or woods just a stone's throw from the drive-thru, and you can squint and see how the vegetation transitions from the typical selection of aggressive weeds to more interesting native or regional plants and trees. This is exactly what my family experienced once after carrying our plastic trays of McDonald's cheeseburgers and french fries and ketchup packets to a few picnic tables erected off the parking lot as an afterthought. We were in a grassy mowed area that abruptly bordered the woods. I sipped my soda (an icy mix of Diet Coke and regular Coke, one of the rebellious pleasures of the self-serve fountain inside the restaurant) and stared into the deepening greenery of the trees just yards away. The seventeen-year cicadas were out that year, and their song of white noise was in full effect, overpowering the whir of cars speeding down the interstate. I saw red raspberries and multiflora rose, staghorn sumac and black walnuts. The pull of the unseeable flora and fauna was magnetic, and for a minute I considered abandoning my cheeseburgers (I always order two) and walking off into the woods, perhaps to vanish from the life of highways and frozen beef patties to exist in a different, parallel plane without physical shape or needs. It was not so much a yearning for escape as an eerie seduction from a slightly sinister source, a bit like the moment in a David Lynch movie where the protagonist succumbs to a nightmarish alternate reality that coexists with our insipidly white-bread one. We were driving home from the airport after weathering twenty-four hours of planes and terminals, and after being in completely artificial holding tanks for so long, my resistance to wild spaces was low.

But I didn't venture out there except to pee behind a tree, my one concession to our unbudgingly untamed essence (the half-half Coke had charmed me more than I'd expected, and I'd consumed it quickly). I finished off my daughter's french fries and helped her assemble her Happy Meal toy, a character from some Pixar movie we'd probably not see in the theaters. Then we gathered up the soiled paper napkins the wind had blown away, threw the whole works into the lonely trash can stationed by the rarely used picnic tables, and climbed back into our car so we could join the other cars on their way to their destinations, civilized and uncivilized.

Raspberry-Glazed Salmon

Serves 3–4

This recipe came about during a visit to relatives in Oregon, which is a great state for berries and salmon. I really like to poke around in people's refrigerators when I stay with them and come up with dishes on the fly. They'd just gone on a fishing trip and then to a U-pick farm, and the hunk of beautiful wild salmon and remaining handful of red raspberries spoke to me.

The bright zip of red raspberries in particular is what works here, while the chipotle powder adds an earthiness that keeps this from tasting dessert-y. It might overpower the glaze at first, but once it's on the fish all of the flavor elements play together harmoniously. You can either broil the fish or grill it; the time it takes to cook will depend on the heat of your grill or broiler, plus the thickness of the fillets.

1 pound (455 g) skin-on wild salmon
    fillets, cut into 4 portions
Salt and freshly ground black pepper
½ cup (60 g) red raspberries
1 teaspoon honey
1 teaspoon apple cider vinegar
A nice pinch of chipotle powder (no
    more than ⅛ teaspoon)

Season the salmon generously with salt and pepper. Set aside.

Meanwhile, prepare the glaze: In a small saucepan, combine the berries, honey, vinegar, and chipotle powder. Set over medium heat, and cook, stirring constantly, until the mixture thickens and the berries break down.

Spoon the glaze over the flesh of the salmon and spread around as evenly as you can.

To broil the salmon: Preheat the broiler. Line a rimmed baking sheet or broiler pan with foil, and lay the fillets on the foil. Slide under the broiler, and cook until the salmon flakes apart, anywhere from 5 to 10 minutes (test the doneness by pressing on the fish gently with your fingertip). Remove the pan from the oven, let cool for a minute, and then use an offset metal spatula to slide between the skin and the flesh—the flesh should release from the skin effortlessly, while the skin remains on the foil. Serve immediately.

To grill the salmon: Prepare a grill for indirect grilling—that is, the heat is on one side of the grill, and the other has no flames or coals. Heat the grill to medium-high.

Arrange the salmon on a sheet of heavy-duty foil. Slide the foil onto the cooler side of the grill, close the lid, and cook until the salmon flakes apart, anywhere from 5 to 10 minutes (test the doneness by pressing on the fish gently with your fingertip). Serve immediately.

SUMMER BERRY FOOL

Serves 8

Berry fool is an old-fashioned English treat whose name derives from the French word *fouler,* to crush. There's no cooking or baking here, and the only way to screw this up is to overwhip the cream. Spoon it into pretty glasses, or serve it as an accompaniment to a thin slice of pound cake or angel food cake.

3 cups (360 g) red or black raspberries

¼ cup (50 g) granulated sugar

3 tablespoons raspberry liqueur, orange liqueur, water, or orange juice

2 cups (480 ml) cold heavy cream

½ cup (60 g) confectioner's sugar

In a large bowl, stir together the raspberries, sugar, and liqueur (or water). Refrigerate for 10 to 15 minutes.

In a chilled bowl, whip the cream and confectioner's sugar to firm but not stiff peaks.

Crush the macerated raspberries with a fork until they reach consistency of a loose jam. Spoon half of the fruit into the cream, and fold once or twice with a rubber spatula; do not overmix. You want a rippled effect. Then add half of the remaining fruit, and fold once or twice. Reserve the remaining fruit puree for garnish.

Divide among serving glasses and serve immediately, or refrigerate up to 24 hours. Just before serving, drizzle the remaining berry mixture over the fool.

ALSO TRY WITH: Currants, thimbleberries, blueberries, huckleberries, or any relatively sweet, juicy, and easily crushable berry.

ROSE HIPS

Rosa spp.
Rosaceae family
Throughout the US and Canada

THE PETALS OF ROSES get all the attention, even in culinary terms—rose petal jam! Quails in rose petal sauce! Rose petals deployed as foofy garnishes! Petals are soft and delicate and immediately identifiable as part of a rose, even to a plant nincompoop. Rose hips are hairy, seedy, tart, and bulbous, and only prominent after the rose itself has long ago spent its showincss. I think of cute boys in high school who unfailingly dated the most boring, listless girls possible: skinny and soft-spoken with long hair and scrunchies that matched their outfits. Those girls were the rose petals. People who prefer more oomph keep their eyes peeled for the hips.

Select your rose hips carefully. Commercially grown roses are notorious targets for generous applications of chemical fertilizers and pesticides. A lot of roses produce scrawny hips, too. You want big, buxom, colorful hips free of harsh chemicals. If rose hips were women, the ones you'd want would look like R. Crumb drew them.

Those rose hips come from what's known as fruiting roses. Rrrow! Sounds foxy! Ramanas rose (*R. rugosa*) is native to Asia but naturalized here. It likes sandy coastal areas but grows inland in sunny spots, too. You'll find it in the Northeast and Midwest, primarily. The blooms are showy and white, pink, or purple with yellow centers—rosy but not too rosy looking. Prairie rose (*R. arkansana*) is a native North

American rose growing in the Plains and Midwest. Their fruits are about ½ inch (1.3 cm) across.

You are not limited to the collection of hips from domesticated roses only, though most likely wild rose hips will be smaller than cultivated ones. Consider that the most plentiful rose hips, those of the multiflora rose, are tiny and tasteless. I see them in the woods in the fall looking like itty-bitty red berries, and I'm tempted to give them a shot, though they barely have any flesh on them.

Rose hips are packed with vitamins. Violet G. Plimmer's *Food Values in Wartime*, published in England in 1941, points out that ¼ ounce (7 g) of rose hips delivers over a third of the amount of the recommended daily dose of vitamin C, and that when imported fruits are scarce, scrappy and resourceful homemakers can still deliver the good to their families. During World War II people collected rose hips en masse to make into syrup or puree so expectant mothers could get their vitamin C.

Some hips are hairier than others. Don't dig in to a fresh rose hip au naturel, as those hairs will irritate your mouth, and possibly your intestines. You can carefully nibble a ripe rose hip, but you'll want to target soft-ripe ones and work around the seeds.

One year in August, my mother and I visited Old Salem, a living history museum in Winston-Salem, North Carolina. A Moravian settlement thrived there for years. The place is fascinating, and I heartily

recommend a visit. After stopping in a number of original structures to hear the spiel of friendly costumed docents, we were about to head out. Then we spotted a giant kitchen garden.

Unlike me, my mom is an excellent gardener, but we're both nutty about plants, so without discussion we made a beeline to the garden and began nosing around. Noticing our interest, the gardener himself came over and started chatting us up. He lived there in Old Salem with his wife, and he tended the gardens using period tools and techniques. This included wearing funny little britches. He was young and friendly and fit, and when he started pulling up some parsnips for us I immediately developed a giant plant nerd crush on him. But the end of his workday was approaching, and it seemed best to get on our way before I embarrassed myself with any more harmless but misled attempts at flirting.

Then I saw them: Growing at the edge of the garden was the showiest cluster of rose hips I'd ever seen! They were bright orange and about the size of olives. "You can take some of those, too," said Farmer Adorable. It was around dinnertime and hot, though, and I opted not to. August rose hips would likely be bland, as exposure to frost develops their flavor, but I regret it to this day; I've not seen a collection of rose hips like that since, though I suspect there was more to it than rose hips. In any case, I'd much rather have a bushel of rose hips than a bouquet of roses.

HONORABLE
MENTION
★ ★ ★ ★ ★

KOUSA DOGWOOD

Cornus kousa
Cornaceae family
Throughout the US in zones 5–8

Bright pink and bumpy, the berries of the kousa dogwood tree are a common sight in the late summer and early fall. Split one open, and inside you'll find pulpy yellow flesh. Visually they are fairly consistent from tree to tree, but their palatability can vary a lot. I love their flavor—it's delicate, a bit like honeysuckle nectar—though their bumpy skins are no pleasure to chew on, and each berry holds precious little pulp.

The tree itself is a common ornamental, one that blooms shortly after its dogwood cousin, *Cornus florida*, also called flowering dogwood. After tasting my first kousa dogwood berry (I recommend splitting one open and sucking at its center before spitting it out), I had visions of making simple syrups from their juice or using the pulp in some kind of chilled dessert. My experiments quickly dampened those high hopes. I tried running the berries through a masticating juicer, which extracted no liquid at all. Then I tried squishing the pulp out of the berries and forcing the berries and skins through a food mill, but the extracted pulp was distractingly gritty. I'm guessing the degree of grittiness perhaps depends on the tree itself, just as the flavor does.

I suspect these are best left as a pleasantly edible novelty to impress your friends during walks, not a lucrative addition to your foraging larder. Should you opt to attempt a kousa dogwood jam or jelly, you'd need to add pectin to get it to set, and be aware that exposure to heat detracts from their delicate, floral flavor. If you are hell-bent on making them into preserves, I'd recommend jelly, since there'd be no gritty solids to contend with.

Successful Failures

It can be frustrating to invest the time and energy in making a recipe that turns out to be a clunker, but the sting is especially pointed when it's made with fruits you carefully foraged. This has happened to me more times than I'd like to recall, particularly when working on this book. When I develop recipes for clients, my job is to remove all possible barriers to success and still have the recipe be somewhat easy to follow.

Foraged fruits can throw wrenches in this. The produce we buy at the grocery stores is bred to be predictably the same whether we're shopping in Boston or San Diego. But foraged fruits can even vary from tree to tree—the thickness of their skins, the firmness of their flesh, their levels of acidity. I've oversalted pickles, undercooked quick breads, and thrown out crab apple chutney because I used apples that were disgustingly mealy. There are some disastrous pawpaw experiments I'd rather not recall at all, pulpy salsas and off-tasting enchilada sauces. The hope and optimism going into these only amplified the lackluster outcome.

I still encourage you to branch out and do your own recipe experiments, but try to take flops in stride. You're going into uncharted territory, a culinary explorer. This happens all the time to chefs in restaurants; we just don't hear about it, because they don't serve flops. When I dump a batch of gross pickles in the compost, or gamely eat a loaf of quick bread that's unappealingly coarse and bland, I think of how the work that went into it was not strictly about the final product. It's about the experience. There's the delightful sense of purpose in setting out with my canvas bag over my shoulder to the crab apple trees I know still have fruit to give me. There are the sounds of the day as I pick them, the people I say hello to as they walk past, the shows on the radio I listen to as I painstakingly halve every frigging crab apple with a little paring knife.

You can't make amazing discoveries without grand mistakes. I just hope you don't waste too many $8 vanilla beans in the process.

Harvesting and Storage

Deadheading roses will lop off their ovaries, and therefore the fruit, so if you're growing your own and you want hips, don't deadhead them.

Rose hips ripen from August through October, but cool weather helps them taste better. Not all rose hips are created equal. Sample before you pick, because some rose hips look fat and lovely but have zero flavor. Pick rose hips after they've attained a deep orange or red and ideally been exposed to a frost. You'll want to get to them before they get too soft. Firm rose hips are tart, and the softer ones with just

a bit of wrinkle are sweeter. (I've heard that freezing and then thawing harvested rose hips can replicate the flavor-building reaction of a frost, but I have not tried it myself.) Every rose has its thorn, so come prepared and either wear gloves or don't make sudden jerky movements.

Refrigerate rose hips after harvesting to keep their nutrient content high. For long-term storage, dry the rose hips as is or freeze them, lopping off the tail and top ends first. A dehydrator dries them faster, but there's a much higher success rate drying rose hips slowly spread out on sheet trays.

Culinary Possibilities

High in pectin, rose hips are great for jams and jellies. You can dry them and use them in herbal tea infusions (pick up most any given box of Celestial Seasonings tea, and there's a 50 percent chance rose hips will be an ingredient).

To make a tea of dried hips, use only 2 teaspoons hips to 1 cup (240 ml) of boiling water and steep for 10 to 15 minutes. Sweeten with honey and enjoy.

Rose hips have very hard, large seeds. Any preparation where they will be strained or forced out—teas, jams, jellies, purees—makes handling them far less time consuming. This includes fresh and dried rose hips. *Nyponsoppa*, rose hip soup, is a Swedish porridge-y fruit dessert made with simmered, sweetened, and pureed rose hips (fresh or dried). The warm soup would have been a thrifty way for winter-bound Swedes to get their vitamins from a forageable item. Today one can purchase instant mixes of the soup. It's traditionally served with almond macaroons.

SERVICEBERRIES

Amelanchier spp.
Rosaceae family
Throughout the US and Canada in zones 4–9

THESE ENCHANTING LITTLE BERRIES have a branding issue because of their name—or names; there are so dang many of them. *Serviceberry* is the least glamorous. One might assume they were so dubbed because enlisted men in a long-ago military campaign ate them out of necessity while suffering in the wilderness. Or that the berries themselves are merely serviceable—not great, not awful. But the berries are wonderful, and their name has an intriguing (though untrue) origin story: the shrubs and trees bloom early in the spring, and folk wisdom went that once the white blossoms appear, the ground should have warmed up enough to allow for digging graves. The service in question would have been a funeral service. This is all bunk, though. In America, "serviceberry" morphed from another name for the plant, *sarvis* or *sarvisberry*.

The berries aren't ready to pick until June or so, the reason behind one of their other names: Juneberries. Native to North America, serviceberries have been naturalized in Europe; in England, they are called snowy mespilus. *Shadberry* and *mountain berry* are, in my opinion, much better names, and you'll find people using those in Rocky Mountain regions. People up north may call them Saskatoon berries (the city in Saskatchewan was named for them). Commercial growers created a Saskatoon Berry Institute of North America, banking on the promise of the berry's powerful nutritional profile (serviceberries rival blueberries for antioxidant and nutrient content).

Serviceberries are attractive shrubs or trees growing up to 25 feet (7.6 m) high. You'll do well to spot them grown as ornamentals in yards on your walks around neighborhoods. In the fall their smallish leaves turn lovely orange and red colors. Often their trunks will be subtly spotted with pale green lichen.

Serviceberries ripen from green to magenta to bluish purple, making a striking array of colors all on the same plant. Their skins have a powdery burnish, just as blueberries do. I eat the darkest ones I can find. You are competing with birds for the berries, so don't put off picking for long, else you'll return to a tree stripped of its lovely fruit.

Also blueberrylike is their flavor, though it's tinged with a grapey vibe, too. They're different enough to be interesting but familiar enough to be welcoming. The skins are tender, and the flesh is juicy and soft. Birds who snack on them have excellent taste. I could eat a whole mess of these, but there aren't enough shrubs in my neighborhood, so I am content to get what I can.

Serviceberries are pomes (like apples or pears) and technically not berries. Just to keep from muddying the waters more, let's not add *servicepomes* to the towering pile of names, though.

Harvesting and Storage

Pick the nice and ripe serviceberries and eat them on the spot; they are fabulous grazing berries. They'll hold up in the fridge for up to a week, or can be frozen, though once thawed they won't hold their shape. If you have many berries, juicing is an option. Or do the opposite and dry them.

Culinary Possibilities

Use serviceberries as you would blueberries, though they are seedier, which can be texturally distracting in baked goods. Jams and jellies are a good idea; you won't need to add pectin. You can also add serviceberries to a shrub, include them in a secondary ferment of kombucha, or juice them and reduce to a syrup.

CASTOR BEAN

Ricinus communis
Euphorbiaceae family
Eastern and southern US

Beneficial castor oil comes from this cool-looking plant, but all parts of the plant itself are toxic. As few as three seeds can be deadly. Ricin, which is extracted from this plant, has played key roles in fictitious and real terrorist plots to poison adversaries. It is one of the most poisonous naturally occurring substances. How can something so dangerous be so useful?

Native to India and Egypt, the castor bean must be handled correctly to give up its benefits. Cold-pressing the bean (really a seed, not a bean) extracts the oil, but not the toxin, ricin, which is not fat soluble. Castor oil has long-documented uses in ayurvedic medicine as a remedy for constipation, and it's long been used topically for healthy skin and hair. (Fun trivia: Castor Oyl was Olive Oyl's older brother in the *Popeye* comic strip.)

People grow castor bean plants because they have a decorative, tropical look and easily attain an impressive height—growing over 8 feet (2.4 m) tall in one season in temperate zones, where they must be grown from seed every year. There was a huge one growing one summer in our local community garden; the person who maintained that plot said she'd read that its presence helped keep mosquitoes away. I was never able to find research to back this up, but its star-shaped leaves towered above the other plants there.

The fruits cluster around a central stalk. They are spiny seedpods, either red or pink, and they hold pretty, mottled seeds—the "beans." The plant can be weedy or invasive, so there's a chance you could run across it outside of a garden or yard, particularly in warmer climates, where it grows as a perennial.

Television fans may know of ricin as a poison Walter White uses twice in the series *Breaking Bad*. It makes a great plot point, but it's a bad idea for some gardeners. Anyone with children or curious pets around should think twice about growing this plant—and, come to think of it, it's not so great to put in a community garden, either, because you can't be sure who might be silly enough to try one of those beans.

SPICEBUSH

Lindera benzoin
Lauraceae family
Eastern and southern US; Ontario

NORTHERN SPICEBUSH IS A WOODSY understory shrub, neither plain nor showy, and its peppery berries are not the sort one snacks on, unless one is a bird. Historically, spicebush was a significant plant in early America. Settlers who were unable to obtain familiar imported spices like cinnamon and allspice made do with spicebush bark and berries. They don't taste anything alike, but we'll get to that later.

Like the pawpaw tree, spicebush likes to grow in the shade and close to streams and creeks. You'll find it in woodsy areas from southeast Canada all the way down to Texas. Usually it's not tall, but it can grow as high as 12 feet (3.7 m) or more. The small berries (about the size of a pinkie fingertip) ripen from green to an attractive red. You'll find them only on the female trees. Spicebush grows in the same conditions as goldenseal and ginseng, two other nifty medicinal plants of the eastern deciduous forest. It's also the namesake of the spicebush swallowtail butterfly, which it commonly plays host to.

Spicebush has glossy, dark green leaves. Crush them with your fingers, and take a whiff—they should smell pleasantly sharp and citrusy. Another way to spot spicebush is to look for its petite clusters of yellow flowers in the spring, a feature that has led spicebush to be called the forsythia of the wilds. In the fall the leaves turn bright

yellow. In the winter, when the shrubs are naked, freckly little whitish bumps along their bark can help you identify them.

The nonfruit parts of spicebush get as much ink as wild edibles as the berries do. The leaves are used to make a pleasant tea, and the twigs can be infused in liquids (like you would a cinnamon stick) to lend them a citronella essence. Native Americans used spicebush berries to make medicinal teas and healing poultices.

Forest critters often eat the berries before you can get to them, so either stake out a tree or just consider yourself lucky to find one. When the berries are brilliant red, pick away, but don't get greedy and strip the entire tree—unlike your forest friends, you can buy food at the grocery store. If you don't have a container, the berries will hold up in your pocket pretty decently until you get home.

As the name implies, spicebush berries are intensely flavored. Often they are described as being similar to allspice, but I think that's way off. Allspice is earthy and dark, while spicebush is sharp and bright. Spicebush is totally its own thing. You'll be most rewarded when you taste it without preconceived notions about it as a foraged understudy for other familiar spices.

There's no harm in eating the berries raw, but it's not the sort of thing one does by the handful. Crush them between your fingers, and you'll notice they are resinous and a bit greasy. When you grind the dried berries, they do not a produce a powder one can easily dust daintily over oatmeal.

For landscaping, spicebush makes a fine native alternative to other fruiting shrubs if you live in the right conditions. Bird-watchers are wise to plant them, as migratory birds like to swoop down on them for snacks and rest. Spicebushes are dioecious—male and female flowers grow on separate plants—so if you're planting any, plant multiple.

Harvesting and Storage

Look for ripe red spicebush berries in the late summer and early fall. Drying them intensifies their flavor tenfold, prolongs their shelf life, and allows you to treat them as their name implies: as spices. They'll get very hard and brown and look like deer poop. I use a dehydrator, but you can leave them out on a windowsill for about 5 days, and they'll handily dry on their own. If you want to play around with the fresh berries, refrigerate or freeze them as soon as you get home; they quickly start to get wrinkly and brown in spots.

For the average home cook, a generous handful of berries should be plenty for a yearlong supply in your spice collection. If you run across a lot of berries and get carried away, perhaps put them in little jars for gifts.

Spicebush berries are very high in oils. I have kept my dried whole berries at cool room temperature without them going rancid, but if you are concerned about it, pop them in the freezer. Once you grind them, though, they can go off pretty quickly, so I recommend only grinding what you think you'll need in a recipe and refrigerating the leftovers for a week or so.

Culinary Possibilities

We're used to using the same vernacular of familiar flavors to describe our food; what I love about spicebush is how it throws curveballs at my expectations. I mess around with it a little more every year, and it still delightfully flummoxes me.

I think spicebush shines brightest when it's regarded not as a stand-in for warming spices like cinnamon and nutmeg. It's piney and bright with a medicinal edge and fruity-hot like black peppercorns—perfect for savory applications, rather than the sweet ones usually described in foraging literature. That resinous sharpness makes it fantastic with rich and fatty grilled meats. Kelly Sauber of Fifth Element Spirits in Shade, Ohio, uses spicebush as one of the aromatics in his gin, which is pretty brilliant, since spicebush and juniper have similar attributes.

If you grind dried spicebush berries in an electric spice grinder, the heat from the friction may render them to a rough paste with the consistency of a crumbly wet rub. And even if you grind them coarsely, your result will have an oily feel, which takes some getting used to. I've seen Dave Rudie, a chef in my area who has a passion for foraged foods, put dried spicebush berries in a pepper mill and grind them fresh over dishes as a finishing element. It seems like a much handier way to use spicebush in small amounts.

Fresh berries are something I have not messed around with much. They are a lot milder than dried ones, which would free you up to pair them with less assertive flavors. You can chop the whole berry up very fine (the seeds are soft when fresh) and add that to sautés, but I find the flavor gets lost this way. It might be interesting to throw a small handful in with shredded cabbage when you get a crock of kraut going, or to add a couple to a stew. I like the idea of gently steeping some in cream on the stove and seeing where that would go: an unusual panna cotta, the sauce for a fricassee. If you get your hands on some, the potential for experimentation is high, because there's not a long, stale list of default spicebush preparations. It's an age-old foodstuff that's practically new to us. What more could a curious forager ask for?

Pattern Recognition

Once you learn how to read, it's pretty impossible to glance at anything with words on it—a road sign, for example—*without* reading it. It's a knee-jerk thing.

Recognizing plants is exactly the same way. After learning to identify key characteristics (distinctive leaf shapes, bark, or blossoms), you'll begin to spot others without much effort. This is one of my favorite things about foraging: Suddenly, trees and shrubs you may have walked past a hundred times will jump out at you, and you'll experience what's out there on an entirely new level. Our eyes delight in zeroing in on what we care about.

It can take a few sightings to get the hang of it. If a plant is off your beaten path, you may only see it once every year or two, and in that span of time memory can play tricks. Don't feel defeated if you need to refer to your guides (whether printed, online, or actual human experts) multiple times before you feel confident in identifying something on your own. In fact, you should refer to guides regularly even when you're positive about a plant's identity. I come away with new pieces of information after reviewing an entry in one of my favorite resources, plus I'm always running across new ones. A foraging enthusiast's library tends to grow in proportion to their experience, and pretty soon you'll be happily cross-referencing new discoveries like the plant nerd you truly are.

Spicebush–Sumac Rub for Grilling

Makes enough for 2–3 pounds (910 g–1.4 kg) of meat

Spicebush is assertive, piney, and citrusy—all qualities that make it great on red meat. This rub holds up to grilling like a champion and helps create an appealing crusty char. The rosemary echoes the pine flavor, the tart sumac adds an edge, and musky white pepper provides an interesting counterpoint to the spicebush. Black pepper, which is fruitier, is fine, too.

I like this on flank steak, rib eyes, strip steak, or any good grilling cut of beef. Try it on leg of lamb or lamb loin chops. One day I'll get my hands on good venison and bust this out—I've read early European settlers used spicebush on wild game. If you don't eat meat but you love spicebush, try this on summer squash or eggplant, mixing in a teensy amount of olive oil to help it adhere.

1 tablespoon freshly ground dried
   spicebush berries
1 teaspoon freshly ground white pepper
1 tablespoon ground sumac
1½ teaspoons kosher salt
1 tablespoon finely chopped
   fresh rosemary

In a small bowl, mix together all of the ingredients with your fingers. It may resemble a crumbly paste, or it could be more powdery—it'll depend on the spicebush.

Pat the meat dry with paper towels, then pat the rub generously over all its surfaces. If you have time, refrigerate the meat with its coating, uncovered, for up to 24 hours (this helps you get a nice crust after grilling).

Any leftover rub will keep, refrigerated, for a few days.

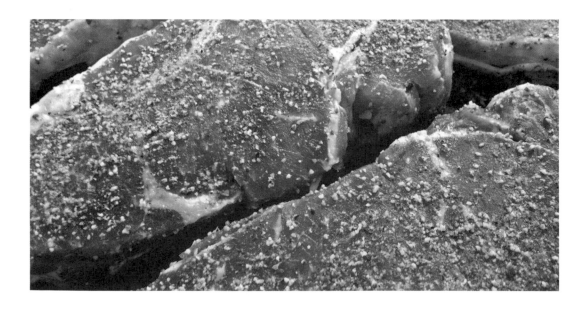

STRAWBERRIES

Fragaria spp.
Rosaceae family
Throughout the US and Canada

AMONG FRUITS, STRAWBERRIES are perhaps the most sexualized. They also convey the most innocence. Why the dichotomy? There's a reason Strawberry Shortcake heads up that franchise of spunky rainbow-haired dolls. But there's also a reason painter Hieronymus Bosch used strawberries as a recurring symbol for the sin of fornication. See the pale, emaciated, naked Flemings grappling miserably over boulder-sized strawberries in *The Garden of Earthly Delights* and you get the sense that eating giant fruit is not only unfulfilling, but something your eternal soul will pay dearly for. (What a drag life must have been in churchy fifteenth-century Dutch circles!)

Wild strawberries are as fleeting as sexual innocence in the face of opportunity. In Roman Polanski's 1979 film adaptation of Thomas Hardy's *Tess of the D'Urbervilles*, the cad Alec D'Urberville dangles a perfect, ripe strawberry before the nubile country maiden Tess. "I would rather take it from my own hand," she says. But Alec accuses her of being coy, and she relents, easing her ruby lips over the fruit and accepting it. After failing to seduce Tess, Alec rapes her, and eventually (spoiler alert) Tess will be hanged for murdering him.

Symbolically, the strawberry straddles the before and after, but in our actual lives there's no tension in the moment itself: Anyone who has come across wild strawberries instantly falls into a carefree

reverie, very much as a teenager making out in a car. Boundaries peel away, though overindulgence in wild strawberries results in a stomachache, at worst. Poor Tess—Maiden No More—eats that single fateful strawberry and emerges from the figurative backseat in the family way.

Humans have eaten strawberries without ruining their reputations probably since they first spotted them, though the earliest records of wild strawberries in human feces date from the Mesolithic era (10,000 to 5,000 BCE). Bosch be damned—fornication is fun, and ideally free of scarring, spiritually or physically. The times we live in are in many ways dark, but our comparative enlightenment about sexual matters should translate without effort to the shameless seeking and consumption of wild strawberries, perhaps as our hunter-gatherer ancestors did.

There are numerous varieties of wild strawberries growing in North America: alpine strawberries or wood strawberries (*F. vesca*, also called *fraises de bois* in Europe and on the pages of *Martha Stewart Living*), though the Virginia strawberry (*F. virginiana*) is more prevalent. The beach strawberry (*F. chiloensis*) grows in dunes and grasslands along the West Coast.

Strawberries are perennial, and they send out runners (stolons in botany-speak) just like their cultivated relatives. We're so used to the sight of strawberries that the seeds on their exteriors don't jar us, but it is unique among fruits. Botanists don't consider strawberries to be true berries, but "aggregate accessory fruits": Each individual strawberry seed is an ovary and is technically a separate fruit.

Wild strawberries are much smaller than those hollow, crunchy, off-season truck farm monstrosities most of us allow to pass for strawberries. Cultivated in-season berries, the kind you pick at local farms, may still dwarf their wild cousins. They will also be much easier to gather a significant amount of, since wild strawberries grow in clusters, but not giant swaths. Look for them from May to October, depending on the climate, in well-drained areas with some sun exposure: hillsides, slopes, patchy forests.

Strawberries are fairly easy to cultivate and are a great deal of pleasure to eat straight from the vine. A gardening friend of mine consumed them that way exclusively, spitting out any specimens she felt had subpar flavor and moving on to the promise of the next berry. This technique can also be employed with wild strawberries, exalting in the moment. Perhaps this basal sort of behavior is the root of the strawberry's role in art, stories, and song as a portal to sensual bliss— indicative of either naïveté or worldliness, depending on how uptight you are. Control of one's manners and bearing are lost upon biting into the delight of a ripe wild strawberry, and I encourage you to pursue this state of being, fleeting as it is.

Harvesting and Storage

Eat these on the spot, as many as possible. If you are fortunate enough to come across a cache of wild strawberries, take some home for people you really love, even if you have to do it using an inverted baseball cap to hold the berries. And for God's sake refrigerate them as soon as possible—wild strawberries have the shelf life of a mayfly.

Culinary Possibilities

As far as dressing them up, little should be required. Some lightly sweetened whipped cream à la Wimbledon would be tasty, as well as an amusing juxtaposition of the highly refined with the savage.

I think cooking wild strawberries is a waste; save that for the flats from the U-pick farm. But if you bring wild strawberries home and their flavor in the shade of your abode underwhelms you because the intoxication of strawberry Kismet has passed, macerate them in a little sugar, or throw them in a shrub or kompot.

To really concentrate the flavor of strawberries, toss them with some sugar, scatter them on a sheet pan, and roast them in a 400°F (200°C) oven until they shrink down and give up some juice. These are really great with yogurt or panna cotta. Instead of a sheet pan, put the strawberries in a baking dish so less liquid evaporates, and this will yield a jammy compote.

Light and Fluffy Old-Fashioned Berry Shortcakes

Makes 9 shortcakes

Biscuit-style shortcakes have a homespun charm, and their relatively low amount of sugar really lets the sweetness of the berries shine. The mixing technique is similar to that for scones, but there's a beaten egg added here for a cakier crumb. Cake flour also helps make these light and fluffy, but regular all-purpose flour works fine if that's that all you have.

For the shortcakes

2 cups (240 g) cake flour
½ teaspoon table salt
4 teaspoons baking powder
½ teaspoon cream of tartar
3 tablespoons granulated sugar
½ cup (1 stick) cold unsalted butter
⅓ cup (80 ml) heavy cream or
    whole milk
1 egg

For the berries

4 cups (580 g) fresh mixed berries,
    such as strawberries, raspberries,
    and blackberries
¼ cup (50 g) granulated sugar, or to
    taste (some berries are sweeter
    than others)
½ teaspoon finely grated lemon or
    lime zest, optional

For the whipped cream

1 cup (240 ml) heavy cream, chilled
2 teaspoons granulated sugar

To make the shortcakes, preheat the oven to 425°F (220°C). Line two baking sheets with parchment paper or silicone baking mats.

In a large mixing bowl, whisk together the flour, salt, baking powder, cream of tartar, and sugar. Grate the butter on the large holes of a box grater, and toss it with the flour mixture. Using your fingertips or a pastry cutter, work the butter into the flour until it looks like fine crumbs. Measure the cream in a glass measuring cup, crack the egg into it, and beat with a fork until well combined.

Pour the cream mixture over the flour-butter mixture, and with your hands, gently work until it comes together to form a rough, shaggy dough that's slightly sticky. (Add a sprinkle of flour if the dough is too loose; add a drizzle of cream if it feels too dry or crumbly.) Knead for four or five turns on a lightly floured surface, and pat into a 7 × 7-inch (18 cm) square. Cut into nine squares, and transfer to the baking sheets (giving the shortcakes plenty of room allows them to brown more evenly).

Bake until golden in spots, 15 to 20 minutes, rotating the baking sheets front to back and top to bottom halfway through. Cool on a rack before serving. The shortcakes may be made a day in advance.

To make the berries, rinse and drain the berries. If you're using strawberries, stem them and halve or slice them. Toss together with the sugar and (if using) lemon or lime zest. Let stand at room temperature for 30 minutes.

Shortly before serving, make the whipped cream. Combine the heavy cream and sugar in a large bowl, and beat by hand or with an electric mixer until it makes soft, rounded peaks.

Not All Returns on Time Spent Foraging Are Edible

Foraging is an activity, but it doesn't have to be high impact. Efficiency is not the point. Think of it as going to nature's spa.

In *Stalking the Wild Asparagus*, Euell Gibbons reflects on what he dubs "the economics of wild strawberries." He once spent an afternoon gathering wild strawberries in an abandoned orchard 20 miles (32 km) away from his house. "The berries were so thick they covered the ground, looking like a red carpet unrolled before me. I could sit down and pick a quart of berries without moving... the day was a revel in beauty, flavor, and aroma, and at its close I felt that I had spent few more worth-while days in my life."

Gibbons's haul for the day was 12 quarts (12 L), an amount a brisk picker can easily bag at a U-pick farm in an hour. He wasn't using quantity as a metric, though. The day was about berries, but also not about berries. Anyone fortunate enough to have enjoyed a transformative taste of a living thing right in the spot where it grew will never forget that experience. If you're lucky, you'll enter an altered state of heightened awareness. Even if your haul is small, you are bringing back much more than fruit.

Gently split the shortcakes with a serrated knife or fork. Spoon the berries and their accumulated juice over the bottom halves of the shortcakes, then top with a generous dollop of whipped cream and, finally, with the top half of the shortcake. Serve immediately, with any remaining berries or whipped cream on the side.

MOCK STRAWBERRIES

Potentilla indica
Rosaceae family
Eastern, southeastern, and western US

Unfortunately, the wild strawberries most abundant in my neck of the woods are weeds growing on lawns. Sometimes called Indian strawberries, mock strawberries, or strawberry grass, they produce tiny berries with all the juice and flavor of Styrofoam pellets. These are not true strawberries, though they are in the same family. Mock strawberries are attractive to small children, who won't be apt to eat any after sampling one. Don't freak—they're not poisonous, just more fun to gather than eat. My daughter likes to pick them and arrange them on the sidewalk, or use as props for fairy food. Lawn maintenance maniacs engage in all-out war with this weed (some states classify it as an invasive), but I find it inoffensive. Anyway, they are not the *Fragaria* you are looking for. These inferior wild berries have yellow flowers and point straight up. Tasty wild strawberries have white flowers and dangle from the stem.

SUMAC

Rhus spp.
Anacardiaceae family
Throughout the US and Canada

"SUMAC? ISN'T THAT POISONOUS?" people will ask if you tell them you've been out and about foraging for the stuff. Great news! Sumac isn't poisonous. You know what is? *Poison* sumac. It's an entirely different plant, though it is in the same family.

That does not mean sumac isn't dangerous in its way . . . at least for me. It's totally safe to eat the crushed berries (which might not appear very berrylike—more on that later), but the dramatic sight of maroon-crowned sumac trees growing on the side of windy highways in the summer and early fall will one day make me wreck my car. Alluring stands distract when I'm driving, and I crane my neck longingly to check them out, neglecting to keep my attention on traffic. Car foraging is worse than texting. I love sumac, but even I would not die for it.

One does not look at a sumac tree and think, "Hey, fruit!" Consider it a spice with sour power. By late summer, when it's ready to collect, you are probably ready for a change of pace from the painstaking picking of soft, fleshy, sweet fruits. Sumac is your remedy. It's a native plant that grows all over Canada and the United States. All edible sumacs have red berries, which is handy to know, but the size and shape of the berries can differ quite a bit.

Sumac is a common sight in fields and open forests. It's a tree you see at the in-between place where the two meet, which is often where

the most interesting things are. The female plants form berries on cone-shaped clusters perched atop their centers. Sumac trees tend to grow in single-sexed colonies. A colony of a dozen or so fruiting female trees in their full fiery glory reminds me of a massive banquet table set with giant candelabras. You can collect a small armload of clusters, process them, and be set up with sumac for a good chunk of the year.

Staghorn sumac (*R. typhina*) grows throughout the eastern United States. It can thrive in dry and rocky soil and on slopes, so maybe that's why I have so many sightings as I speed down county roads. The trees can get as high as 30 feet (9 m). Smooth sumac (*R. glabra* L.), also known as scarlet sumac, does not grow as tall as staghorn sumac. You can tell the difference between the two because smooth sumac has smooth branches, and staghorn sumac has hairy ones like the velvet on a stag's horns (both species have small, furry berries, though). Smooth sumac is the only shrub or tree native to all forty-eight contiguous US states.

As the leaves shrivel and drop, sumac clusters will remain, their color fading, yet still serving as reminders of the season that once was. Birds eat them throughout the winter. Sometimes you'll even see last year's clusters near new clusters. In the spring the flowers start out white, and then become those flaming red berries all over again.

R. coriaria (sometimes called Sicilian sumac) is the sumac you might know from Middle Eastern food markets. It gets sprinkled on hummus, tossed with lemon juice and olive oil in the Lebanese pita salad fattoush, and is smashing on anything with feta cheese. Those sumac berries are more crushable than staghorn or smooth sumac, which have a delicate texture.

Sumac is fruity and tart, but not lemony. The default sumac preparation is sumac-ade, which Native Americans would make by simply

steeping sumac clusters in cold water. The sumac tints the water crimson pink, and you can sweeten it a tinge for a light, cooling summer drink. Boiling sumac makes it tannic and bitter, so cold infusions are the way to go.

Oh, and about that poison sumac. It's *Toxicodendron vernix* and grows in swampy areas in the eastern United States. Luckily, you are more likely to run into the prettier, un-nasty sumac. Poison sumac has clusters of white berries jutting out from its main stem. Don't touch any part of it! The sap contains urushiol, the same compound in poison ivy and poison oak that gives us awful rashes. Admire it from a distance. As for edible sumac, you should cozy up to those trees. Just don't let them get you in a car wreck.

Harvesting and Storage

Bring your pruning shears. Better yet, bring cut-and-hold extendable pruning shears. The cluster crowning a tree can be way up there. I have stood on inverted 5-gallon buckets on tiptoe with my arms stretched to the max trying to zero my pruners in on the stem below the cluster. Wear gloves, if you like, because parts of sumac trees are sticky.

Look for the brightest clusters. As the weeks pass, their saturated tone starts getting murkier—along with their potency. Rain will lessen it, too. Wet weather as summer turns to fall is not good news for sumac hunters. You can even lick your finger after touching a cluster and get a sense of the flavors it will soon lend to your cooking. You want clusters that are noticeably sour. Stick the clusters you've lopped off in a paper grocery sack. They can hang out in the bag for a few days, in fact.

Pick off any funny-looking spots and then rinse the clusters off briefly (you don't want to wash away the flavor). If you'd like to infuse the sumac in water, just keep it on the cluster.

If you'd like to make a dry spice, break the berries away from their center in bunches. Dehydrate until dry and brittle, or lay them on a rimmed baking sheet and dry in a very low oven (150°F/ 65°C or so) for 2 to 3 hours.

Put the dried berries in an electric spice grinder or food processor. You're trying to liberate the powder from the seeds, which are hard and not tasty. Use a sieve to sift out the red powder and store in an airtight jar. I say toss out the seeds, though some people keep them in. Sumac will lose its potency over time, but it'll still be tart even after a year.

Culinary Possibilities

You can't quite swap out our native sumac one-to-one for use in Middle Eastern recipes, since it's subtler. It's also fluffy, so a tablespoon of it would weigh less. But you can adapt and play around. I like it best as a finishing element. It'll make a kick-ass fattoush and a flashy topper on deviled eggs, hummus, or flatbreads. Sumac works best with poultry, fish, legumes and beans, and vegetables. But that's a pretty formidable list!

In *Stalking the Wild Asparagus*, Euell Gibbons wrote of being short on lemons when he had elderberries to cook into jelly, so he cooked the elderberries in sumac-infused water. "It had a delightful, clean tartness that invited you to eat more, and then more." The tartness of sumac *is* clean. It wakes up your taste buds without overwhelming them.

Sumac-Ade

Makes about 2 quarts (2 L)

It seems impossible that leaving giant sticky-fuzzy clusters of sumac berries in water would yield anything remotely drinkable. But sumac-ade is eminently drinkable, especially on a hot day. It's tart and bracing, but not harsh. There's a little hint of tannic grip, too. It's a fantastic nonalcoholic aperitif. You could probably use it as a base for neato cocktails, but I like it as is—cold, no ice, no booze, no fizzy water. I'd take this over pink lemonade any day. It's quite different, actually, with an intriguing sourness that's not citrusy at all.

8 cups (2 L) cool, filtered water
6–8 large heads staghorn sumac
2–3 tablespoons light agave nectar

Put the water in a large pitcher or glass jar. Pull the sumac berries off the heads so you have clusters that will fit in the pitcher (this is messy, so you may want to do this outside). Swish them around, and then either let the whole works steep at room temperature for 2 to 4 hours or (my preference) refrigerate overnight.

Line a strainer with cheesecloth or a clean piece of an old T-shirt (the cloth will get stained). Set over a bowl, and pour the infused sumac though it. It should pass through pretty quickly, and there won't be any fluttery sumac residue in the liquid, surprisingly.

Transfer the sumac-ade to a pitcher, and stir in the agave nectar a tablespoon at a time, tasting as you go. You want a hint of sweetness, but not much. The sumac-ade tastes best a day after it's mixed with the agave nectar, which at first overpowers the drink but then mellows and creates a nice harmony. Refrigerated, the sumac-ade will keep for quite a while, but I doubt it will be around for more than a few days.

Za'atar

Makes about ⅓ cup (80 ml)

The Middle Eastern spice blend za'atar is perhaps more of a concept than a recipe—the formulas for store-bought versions vary quite a bit. I've seen ones that are very heavy on the sesame seeds, and others that are mostly dried herbs.

When I use za'atar, I use a lot of it. Sprinkle it generously over fried eggs, on rich plain yogurt topped with chopped tomatoes and cucumbers, on a bowl of hummus, or—my favorite—on cubed potatoes before roasting the hell out of them.

¼ cup (11 g) ground staghorn or
    scarlet sumac
2 tablespoons dried thyme
2 tablespoons dried oregano leaves,
    or 1 tablespoon marjoram plus
    1 tablespoon oregano
1 tablespoon roasted sesame seeds

Combine all of the ingredients in a mortar and pestle, and mash them up until the herbs are broken down somewhat and the aroma of the herbs and sesame seeds is fragrant.

Transfer to a clean jar, label, and date. Za'atar will keep for 1 year but will be most flavorful if used within a few months.

Za'atar Roasted Potatoes

Serves 4

I had french fries dusted with za'atar at a fancypants gastropub once and was hooked. I like the combination better roasted, because the spice mixture fuses better with the cut surfaces of the potatoes and you get more flavor in every bite.

1½ pounds (680 g) Yukon gold potatoes, cut into 1-inch (2.5 cm) chunks
2 tablespoons olive oil
1 tablespoon za'atar (page 350)
Kosher salt to taste

Preheat the oven to 450°F (230°C). If you have a convection setting, use it.

Spread the potatoes on a rimmed baking sheet. Drizzle with the olive oil, sprinkle on the za'atar, and season generously with salt. Toss it all together so the potatoes are well coated. They should be quite freckly with the za'atar.

Roast in the oven to 15 to 20 minutes, then shake the pan and roast until the edges of the potatoes are nice and brown and crispy, about 15 minutes longer. Usually I want to shove one of these in my mouth straightaway, but wait until they have cooled a minute or two.

ADDITIONAL RECIPES

SOMETIMES WHAT YOU REALLY NEED isn't a recipe for fruit, but a recipe for something to go with fruit. That's what these are: things to make to accompany or use up your stocks of jams, dried fruits, exemplary fresh fruits, or odds and ends of frozen fruit.

Kompot (Eastern European Fruit Punch)

Makes 2–3 quarts (2–3 L)

Call kompot the Kool-Aid of Russia. Imagine relying on questionable drinking water long ago, and now imagine making the water safe and flavorful by simmering it with fruit and some sugar. It's cool and refreshing, and it comes in as many varieties as your imagination. It's also fabulous to make when you have an assortment of fruits that are nearly, but not quite, over the hill.

Russian émigrée Margarita Gokun Silver wrote about kompot in her essay "Cold Water and Roman Pasta":

> *In my family kompot was a joint effort. My grandfather, my mother, and I harvested the fruits that grew in abundance at our dacha and my grandmother made them into kompot. She boiled combinations of gooseberries, strawberries, raspberries, currants, pears, and apples in water and sugar long enough for the liquid to assume the color of the darkest fruit in the mixture. She then served it warm or at room temperature while leaving enough to preserve several five-liter jars for the winter.*

2 handfuls strawberries, hulled and cut in half

2 handfuls cherries, stemmed and cut in half (no need to pit)

1 stalk rhubarb, cut into 1-inch (2.5 cm) pieces

1 handful blueberries

1 handful grapes (white or red), halved

1 apple, cored and sliced

1 orange, cut into 1-inch (2.5 cm) wedges

8 cups (2 L) filtered water

Up to ¼ cup (50 g) granulated sugar, *or* a few tablespoons of honey or agave nectar

Combine all the ingredients except the sugar in a nonreactive stockpot, and bring to a boil. Cover, reduce the heat to a simmer, and cook gently for 20 minutes. Turn off the heat, and let the pot sit uncovered for a few hours to steep.

Pour the kompot through a sieve into a pitcher or large jar; discard the fruit (it'll be mushy and unattractive at this point). Add sugar, honey, or agave syrup to taste, then chill. Taste again before serving, and adjust the sweetness levels, if needed. Serve chilled, with or without ice, and garnish with a little fresh fruit in each glass, if you like.

NOTE: Mushy-ripe fruits like bananas, medlars, pawpaws, and Hachiya or American persimmons are not good in kompot. Don't go too crazy with citrus, either.

ALSO TRY WITH: Use any stone fruit you like—apricots, plums, peaches, nectarines. Pears and Asian pears are a fine addition, as are loquats, elderberries, mulberries, currants, and nearly any kind of caneberry. Also don't discount tart wild berries like salal or chokecherries, but do use them in moderation.

Rich Rice Pudding

Serves 4–6

There are baked rice puddings and stovetop rice puddings. The latter are creamier, easier to make, and a splendid canvas for fresh or cooked fruit. If you're at a loss for what to do with all those foraged fruit compotes, syrups, and reductions, here is your solution. Gently poached or just plain fresh fruit is just as good with the pudding.

6 tablespoons arborio rice
¾ cup (180 ml) milk
2 cups (480 ml) half-and-half
½ cup (100 g) granulated sugar
Pinch salt
Finely grated zest of ½ lemon
1 (2 to 3 inch/5 to 7.5 cm)
    stick cinnamon
1 egg yolk
1 teaspoon vanilla extract
1 tablespoon unsalted butter

In a medium heavy-bottomed saucepan, combine the rice and milk. Bring to a boil, reduce to a simmer, and cook partially covered until the milk is absorbed, 6 to 8 minutes.

Stir in the half-and-half, sugar, salt, lemon zest, and cinnamon stick. Bring to a boil, reduce the heat to a simmer, and cook, stirring frequently, until the rice is tender, 20 to 25 minutes. Over low heat, stir in the egg yolk and cook, stirring constantly, for 1 minute. Remove from the heat.

Add the vanilla and butter; remove and discard the cinnamon stick. Stir the pudding, and let it sit 20 minutes, during which time it will thicken to a very nice consistency. Spoon into serving dishes to serve warm with the fruit accompaniment of your choice. Or chill the pudding; it may absorb liquid as it sits. If so, stir in some milk or water to loosen it up.

SUGGESTED ACCOMPANIMENTS: Balsamic Blackberry Compote (page 76), Roasted Maple Blueberries (page 82), Poached Quince (page 314), Pear Butter (page 272), or any thick but not stiff jam. Served with poached pears, rice pudding is known as pear condé.

Buttermilk Panna Cotta
Serves 8

Light enough to serve at the end of an elaborate meal yet creamy enough to feel dessert-y, panna cotta has just enough richness to offset fruit but not overpower it. These individual custards are set with gelatin and take very little work to prepare.

2½ teaspoons unflavored gelatin
¼ cup (60 ml) cool water
1¼ cups (300 ml) milk
⅓ cup (65 g) granulated sugar
1 vanilla bean, halved lengthwise
2 cups (480 ml) full-fat buttermilk

Sprinkle the gelatin over the water, and set aside for 10 minutes to hydrate.

Combine the milk and sugar in a medium saucepan. Scrape the seeds from the vanilla bean with a paring knife, and add them to the pan, along with the bean itself. Bring to a boil over medium-high heat, stirring constantly, until the sugar dissolves. Remove from the heat, and let steep, covered, for 10 minutes. Remove the vanilla bean. Add the gelatin mixture, and stir until it's fully dissolved (return the pan to very low heat, if needed). Stir in the buttermilk.

Divide the mixture among eight 6-ounce (180 ml) ramekins or custard cups. Refrigerate, covered, until set, at least 4 hours. The panna cottas may be made up to 2 days in advance. Serve with the fruit accompaniment of your choice. Balsamic Blackberry Compote (page 76) is nice.

Sweet Vanilla Mascarpone Cream
Serves 6

When amazing fresh fruits need a little extra something to elevate them to a dinner party dessert, whip up a batch of this.

1 cup (240 ml) mascarpone cheese
2–4 tablespoons confectioner's sugar
½ teaspoon vanilla extract
2 tablespoons heavy cream
Your choice of fresh or macerated fruit, or fruit compote

In a large bowl, beat the mascarpone until smooth. Add 2 tablespoons of the confectioner's sugar along with the vanilla and cream; beat until somewhat lightened (it won't be fluffy like whipped cream, just a little lighter than when you started). Taste, and add more powdered sugar, if you like. Serve over the fruit, or serve the fruit on top.

Rye Pastry

Makes enough for 2 crusts, or 1 double-crusted pie

We think of the flavor of rye as the flavor of caraway seeds, because caraway seeds are always part of deli rye bread. Rye on its own is slightly fruity, though. A rye crust has a magical complexity when paired with fruit pies, particularly berry, apple, or pear pies. It smells incredible as it bakes, both familiar and exotic.

My favorite liquid to use in this is cheap pilsner beer, preferably a very cold can of Hamm's—some for the dough, and the rest for me. I'm not sure if it makes much of a difference in the final baked pastry, but gunking up the Hamm's can with doughy fingers is really fun.

Don't use coarsely ground rye flour here; it'll make the dough dry and bake up with bothersome grainy bits. I like to get rye flour from bulk food bins, because it's fresher and I see for myself how finely it's ground.

Scant ⅔ cup (75 g) fine rye flour

1½ cups (195 g) unbleached all-purpose flour

¾ teaspoon table salt

1 teaspoon granulated sugar

1 cup (2 sticks) cold unsalted butter

About ⅓ cup (80 ml) ice-cold pilsner beer, apple juice, kombucha, or water with 1 teaspoon apple cider vinegar added

On a clean countertop, set out a piece of plastic wrap about a foot and a half (45 cm) long.

Mix the flours, salt, and sugar in a large bowl. Set aside.

Cube the butter into ¼-inch (6 mm) chunks, or—this is my preference—quickly grate it using the coarse holes of a box grater. If the butter is starting to get soft, pop it in the freezer for 10 minutes or so before proceeding.

Add the butter bits to the bowl of dry ingredients, and toss to coat. With the tips of your fingers, smash and pinch the tiny butter cubes to make crumbles and flakes no larger than peas. Stick the bowl in the freezer for a few minutes if the butter starts to feel greasy.

Sprinkle about 3 tablespoons of liquid over the mixture, and toss it all together like a salad. Continue tossing and adding liquid, 1 tablespoon at a time, until everything comes together to form a rough, shaggy mass that holds its shape when you press a clump together; err on the side of slightly sticky, rather than dry and crumbly, if you're unsure when to stop. Shape into a ball, divide in half, place the halves on the precut plastic wrap, and flatten into disks about 6 inches (15 cm) across. Wrap each securely, and chill in the refrigerator for at least half an hour, and up to 2 days, before rolling. Dough may also be frozen for up to 6 months.

NOTE: I like to take any trimmings left over from rolling out a piecrust, form them into a ball, and chill them for 5 or 10 minutes. Then I roll it into a rectangle, sprinkle it with cinnamon sugar, and fold it over before rolling to a ⅛-inch (3 mm) thickness. Cut into strips about ¼ to ½ inch (6 to 13 mm) across, sprinkle with more cinnamon sugar, and bake at about 400°F (200°C) until crispy and golden brown. What a treat!

Crème Fraîche

Makes 1 cup (240 ml)

There's heavy cream, and there's crème fraîche—which is simply cultured heavy cream, but it's thicker and tangier. You can whip it as you would heavy cream to serve with desserts. It won't fluff up as much, but it sets fruit off better than plain whipped cream.

Ultrapasteurized cream does not work as well for this. The ambient temperature of your kitchen seems to make a difference in how long it will take for the cream to set, too. I've had it take up to 24 hours to thicken, and I've had it set in under 2 hours.

1 cup (240 ml) heavy cream
2 teaspoons white vinegar

Mix the cream and vinegar in a glass measuring cup or jar, and allow to sit at room temperature for 2 hours, or until the cream is thickened and tangy. Refrigerate until ready to use. Crème fraîche will keep, covered and refrigerated, for 2 to 4 weeks.

Stir up the crème fraîche to loosen it, and serve as is as a garnish for fruit desserts. If you prefer, you may sweeten it with a little powdered sugar. To whip crème fraîche as you would fresh heavy cream, whisk it in a chilled bowl until it's lighter in volume and makes soft peaks when you lift up the whisk (it won't be as voluminous as whipped cream). Or thin crème fraîche with a little milk or water, and drizzle over soups as a garnish.

Fruit-Friendly Muesli

Makes 5 cups (1.2 L), enough for 10 servings

I eat muesli in the summertime when I want to kick off a productive day of getting dirty and sweating a lot. Muesli is different from granola in that muesli isn't baked or sweetened. If you have been drying a lot of fruit, add it to this. An overnight soak helps your body better absorb the nutrients, and it improves the texture so you feel you're having a treat and not a horsey feed bag.

1 cup (100 g) rolled oats
1 cup (125 g) rye flakes
1 cup (125 g) wheat or barley flakes
½ cup (70 g) mixed seeds (sesame, pumpkin, flax, or chia)
¾ cup (90 g) chopped mixed nuts (almonds, hazelnuts, walnuts, or pecans)
2 tablespoons cacao nibs, optional
1 cup assorted dried fruit
⅛ teaspoon fine salt

In a large bowl, combine all of the ingredients, and toss together with your hands, breaking up any clumps of dried fruit, if necessary. Transfer to a large jar or container with a tight-fitting lid, and keep at room temperature for up to 2 months.

The night before you plan to eat breakfast, put ½ cup muesli in a bowl and stir in ½ cup (120 ml) yogurt, milk, buttermilk, cream, or the nondairy milk of your choice. Cover, and refrigerate. In the morning, stir, add a little extra liquid, if needed, and top with some fresh fruit. Half a grated or chopped apple or pear adds a nice crunch. After eating, plop your empty bowl in the sink and run out and scale some alpine peaks.

WHOOPS! Forget to get it ready the night before? In a pinch, put ½ cup muesli in a heatproof bowl and add ¼ cup (60 ml) boiling water. Let sit for 5 minutes to soften the grains, then top off with the dairy or dairylike liquid of your choice.

NOTE: To deepen the flavor of the grains before mixing your master muesli, spread them on a rimmed baking sheet in a 350°F (175°C) oven, and bake for 6 to 8 minutes, until aromatic. Transfer to a large bowl to cool before adding the seeds, nuts, and dried fruit.

Toasty Granola

Makes 6–7 cups (1.5–1.7 L)

This is the kind of granola you eat for a wonderful breakfast and not the kind you eat straight from the jar in clusters. That kind of granola is fine, but it's a bit much for me to start the day off with. I don't add dried fruit to batches of granola because in my house, we all prefer different dried fruits, and sometimes I like fresh fruit on there instead.

4 cups (400 g) rolled oats

½ cup (40 g) unsweetened
    coconut flakes

¼ cup (35 g) flaxseeds

¼ cup (35 g) sesame seeds

½ cup (60 g) sliced almonds

Pinch kosher salt

⅓ cup (80 g) melted coconut oil

½ cup (120 g) maple syrup, honey,
    or sorghum molasses

Up to ¼ cup (60 ml) applesauce,
    optional

Preheat the oven to 350°F (175°C).

Line a rimmed baking sheet with parchment paper, leaving a slight overhang on all the edges.

In a large bowl, toss together the oats, coconut, seeds, almonds, and salt. Pour the oil and whatever sweetener you're using over the bowl, and toss to coat. If you like your granola in clumps, add a dab of applesauce, too.

Spread the sticky mixture on the pan in an even layer. This is easier if you lightly dampen your palms with water to keep the granola from sticking to them. Put the pan in the oven, and reduce the temperature to 250°F (120°C). After 30 minutes, stir the granola around. Bake for 30 minutes longer, then turn off the oven and leave the pan in there for 4 hours or overnight. (Put a note on the oven door so you don't forget the pan is in there.)

Break the granola into small bits with your hands, if that's how you like it. Lift the parchment with the granola off the baking sheet and use it as a funnel to feed it into an airtight container. Covered, the granola will keep for 2 months.

NOTE: Granola is all about improvising. I keep the ratios of grains, nuts, oil, and sweetener about the same, but I usually switch up what goes in there based on what I have. Sometimes, if we have a nearly empty box of dry cereal like Rice Chex or Rice Krispies, I'll toss that in, too.

Kolachi

Makes 4–5 dozen

When you make a batch of foraged jam, those jars are precious. They can start to accumulate if you do a lot of foraging and a lot of preserving, though. These cookies are worth cracking those jars open.

We made giant tins of kolachi every Christmas when I was growing up. The recipe comes from my mother's childhood neighbor, Mrs. Johnson, whose background was Polish. These buttery and flaky bundles are like tiny pies. In northeast Ohio *kolachi* can refer to a number of different eastern European baked goods, including nut rolls and yeasted buns. It depends on the particular community. Sometimes cookies like these are called kiffles. I don't care what you call them, but you should make them. Use some of your amazing foraged jams for filling.

2½ cups (325 g) all-purpose flour
1 teaspoon baking powder
½ teaspoon salt
1 cup (2 sticks) unsalted butter, softened
2¼ cups (450 g) sugar, divided
2 eggs
1¼ cups (300 ml) thick fruit jam or very thick fruit butter
1 egg white
1 teaspoon water

In a medium bowl, whisk the flour, baking powder, and salt to combine. In a large bowl, beat the butter with an electric mixer until creamy. Add ¼ cup (50 g) of the sugar and beat until fluffy. Add the eggs, one at a time (the mixture will look curdled). Add the flour mixture, and beat until combined.

Place the dough on a large piece of plastic wrap, and form it into rectangle. Wrap, and refrigerate 2 hours, or overnight.

Preheat the oven to 350°F (175°C). Line two baking sheets with parchment. Place the remaining 2 cups of sugar in a shallow bowl. Divide the dough in quarters. On a floured surface with a floured pin, roll the dough into a square ⅛ inch (3 mm) thick. Cut it into 2½-inch (6.3 cm) squares with a paring knife or pastry wheel.

Put the fruit filling in a quart zip-top bag, and snip off the corner. Pipe about 1 teaspoon filling on each square. Bring the opposite corners in, and pinch to seal.

Beat the egg white with 1 teaspoon water; brush over the formed cookies, then roll them in the sugar. Place the cookies 2 inches apart on the baking sheet. Bake until the kolachi are lightly browned at the edges, 8 to 12 minutes. Remove to wire racks with a spatula, and cool. Covered in an airtight container, the kolachi will keep 1 week. These freeze well.

NOTE: This is a soft dough. It's easier to roll when it's slightly cool. You can reroll the scraps once to get a total yield of 5 dozen; with no rerolling the yield is 4 dozen.

ACKNOWLEDGMENTS

CARRIE HAVRANEK CAME UP WITH THE IDEA for this book. She is full of good ideas, and I'm glad I listened to that one. Carrie, I am so happy to call you a friend.

A couple of years ago, I found out this guy Andy Moore was writing a book about pawpaws, so I emailed him out of the blue, and he generously took the time to talk to me about our shared interest in this extraordinary fruit. This book is in part an outgrowth of that conversation.

Shortly after moving to Marietta, I happened upon a network of mixed-use trails weaving through different woodsy spots in town. They are easy to get to, and I have spent some many contented hours exploring them. The River Valley Mountain Bike Association built and maintains those trails, and without them I don't think I would have developed this foraging habit that's so dear to me. Every time I walk on those trails, I am thankful for the RVMBA's hard work.

One person can write a book, but you need many people to make a book. I've enjoyed working with Chelsea Green Publishing on this project and am especially thankful for the guidance and support of my patient editor, Michael Metivier. Thank God he has a great sense of humor.

Matt Hart, Laura Veirs, Leena Trivedi-Grenier, Karen Solomon, Marisa McClellan, Lynne Curry, Cheryl Sternman Rule, Nancie McDermott, Samara Linnell, Melanie Tienter, and probably a few other people I am forgetting graciously shared their ideas, words, or time with me.

My husband, Joe, and daughter, Frances, are understanding of my periodic need to take off on long walks by myself. I am glad we are all okay doing our own things, apart and together. My parents, Jim and Carol Bir, raised me to feel confident and free out of doors, mostly by just being that way themselves.

Karen, Zsofi, Sarah, Powell, Nia, and Mer: Just knowing I have you to hear me out and give me honest, canny feedback keeps me moving forward. I trust and love you guys.

People who care about weird plants are unfailingly receptive to the clumsy enthusiasm of aspiring fruit nerds. I never feel judged for what I don't know when I'm talking to plant people. Every day, I try to be more like them.

RECOMMENDED RESOURCES

Plant Identification and Care

USDA Plants Database
plants.usda.gov/about_plants.html
A solid first stop for people seeking information
about native or naturalized plants of the United
States and its territories. Many of the species have
detailed Fact Sheets and Plant Guides. The interface
is bare bones, but take a minute or two to click
around and you'll notice resources galore: distribu-
tion maps, classification reports, and invasive status.

**Lady Bird Johnson Wildflower Center
Native Plants Database**
www.wildflower.org/plants
A listing of over eighty-eight hundred plants native
to the United States. Most entries include multiple
high-quality images that are helpful in identification.
They have state-by-state lists of recommended native
species for planting (and Canadian provinces, too).

Missouri Botanical Garden Plant Finder
www.missouribotanicalgarden.org/plantfinder
/plantfindersearch.aspx
Information on and photos of the over seventy-five
hundred plants that are growing or have been
grown in the Kemper Center for Home Gardening
display gardens. The Plant Finder covers plenty of
unusual cultivated fruits (such as medlars) along
with wild ones (many species of sumac).

Hansen's Northwest Native Plant Database
http://www.nwplants.com
A folksy and image-rich list of plants native to northwestern North America, including many food plants.

Calflora
www.calflora.org
A database of plants (native and introduced) growing
wild in the state of California, complete with maps.

Floridata Plant Encyclopedia
floridata.com/plantlist
An online plant database and encyclopedia offering
profiles of cultivated and noncultivated plants from
all continents and climates.

California Rare Fruit Growers, Inc., Fruit Facts
crfg.org/wiki/fruit
The largest amateur fruit-growing organization in
the world (their chapters are not all in California).
The CRFG Fruit Facts wiki contains detailed and
well-rounded information on many cultivated
fruits, much of it from members' own experience.

Eat The Weeds
www.eattheweeds.com
The website of foraging educator and author Green
Deane, who is best known for his in-depth foraging
videos on YouTube. Hundreds of archived articles
on his site offer spirited and informative takes on
many edible plants, fruit and otherwise.

Foraging Texas
www.foragingtexas.com
The site of foraging educator and research chemist
Mark "Merriwether" Vorderbruggen has profiles
of over 150 edible wild plants of Texas and
the Southwest.

North American Fruit Explorers

www.nafex.org

A network of fruit growers and enthusiasts throughout the United States and Canada devoted to the discovery, cultivation, and appreciation of superior varieties of fruits and nuts.

Fruit Preservation

Pick Your Own

pickyourown.org

Listings of U-pick farms all over the world, but also information on how to harvest and preserve common cultivated fruits.

National Center for Home Food Preservation

nchfp.uga.edu

With abundant free e-publications, videos, and online tutorials, this is the most authoritative location on the web about canning, freezing, drying, curing, and more. This is not a place to discover cutting-edge information, but a resource for tried-and-true methods and a valuable reference for food safety practices.

Food in Jars

foodinjars.com

Writer and canning teacher Marissa McClellan has an upbeat tone and a strong eye for detail, making her blog, *Food in Jars*, both trustworthy and readable. Her posts address not just the *hows* of canning, but also the all-important *whys*, encouraging you to learn and grow from your own experiences.

Wild Fermentation

www.wildfermentation.com

The site of Sandor Katz, the Pied Piper of fermentation. Besides his own posts, the user forum here is very active, offering connections and community support.

Miscellaneous

The Cooperative Extension System

www.extension.org

The hundreds of regional extension services operated through the US Land-Grant University System are time-tested resources for agriculture and nutrition-related topics such as gardening and food preservation. A collaboration through federal, state, and local governments, the many state extension offices and websites offer classes and information tailored to the needs of local communities.

Fallen Fruit

fallenfruit.org

A contemporary art collective that uses fruit trees as a jumping-off point to strengthen communities and consider the boundaries between public and private spaces. Projects include site-specific installations, community fruit parks, and the Endless Orchard, a "massive living public artwork" that includes an interactive map to share the locations of public fruit trees, plus encouragement to plant ones yourself.

Falling Fruit

fallingfruit.org

An online tool (not affiliated with Fallen Fruit) built by and for foragers that allows users to "map the urban harvest." The site includes a sharing page with food banks that accept produce donations.

Not Far from the Tree

notfarfromthetree.org

This Toronto-based nonprofit networks area volunteers to pick fruits and share the abundance with homeowners, volunteers, and agencies who can distribute it to those in need.

IMAGE CREDITS

Page vi	Mary (aenigmatēs on Flickr)	Page 114	Liz West
Page x	Virginia State Parks	Page 116	Danny Shortall
Page 5	DC Central Kitchen	Page 118	Glenn Fleishman
Page 13	Linda	Page 120, *top*	Edsel Little
Page 16	Chad Versace	Page 123	Isabell Schultz
Page 22	Samara Linnell	Page 126	Daniel Jolivet
Page 31	Melanie Tienter	Page 130	Anthony Masi
Page 36	YoLaGringo	Page 132	Sevi Moby
Page 39	Samara Linnell	Page 140	Jason Hollinger
Page 43, *left*	Adam Buzzo	Page 142	Leigh Graves Wolf
Page 43, *right*	bkkm/iStock.com	Page 149	Kate Ter Haar
Page 46	Kathy Kimpel	Page 150	Luke
Pages 48 and 51	psrobin	Page 159	K M
Page 56	Wendy Cutler	Page 160	karendotcom127
Page 60	Wendell Smith	Page 163	Quinn Dombrowski
Page 61	F. D. Richards	Page 166	Sten / Wikimedia Commons
Page 64	Ken Ratcliff	Page 172	Peter Stevens
Page 65, *left*	Botanicus	Page 173	Laurel F.
Page 66, *left*	Anne Heathen	Page 179, *top*	National Park Service / Brad Sutton
Page 66, *right*	Drew Avery	Page 180	Old Book Illustrations
Page 67	F. D. Richards	Page 181	Henry Hemming
Page 71	University of Delaware Carvel REC	Page 184	Nancy
Page 75	Andi Roberts	Page 185	Stacy Spensley
Page 76	Adele Prince	Page 186, *top*	Mark Bonica
Page 78	Richard Wood	Page 196	Jon Bragg
Page 84	Melanie Tienter	Page 197	Cayobo
Page 86	Ketzirah Lesser and Art Drauglis	Page 202	Forest and Kim Starr
Page 87	Dave Bonta	Page 203	Yun Huang Yong
Page 88	Tom Brandt	Page 208, *bottom*	Bri Weldon
Page 98	Patti Niehoff	Page 211	Peter Stevens
Page 106	Andrea Pokrzywinski	Page 212	DELTA Intkey
Page 107	Kerry	Page 214	KaCee Holt
Page 108	Liz West	Page 217	Wonderlane
Page 111	InAweofGod'sCreation	Page 232, *left*	Sarah Sammis

INDEX

cranberries (*Vaccinium* spp.),
79, 105–12
in chutney with crab
applies, 101
dried, 20, 108
juice of, 106, 121
in ketchup, 110
medicinal uses of, 106, 121
in muffins, multigrain
multi-berry, 225
pectin in, 108, 113
in pemmican, 80
in relish, 108, 109
in streusel bars, 112
Crataegus spp. (mayhaws), 6,
193, 213–20
Crataegus aestivalis (eastern
mayhaw), 214
Crataegus opaca (western
mayhaw), 214
Crème Fraîche, 357
crocks, earthenware, 21
curd, citrus, 192, 252
with pawpaw, 252
currants (*Ribes* spp.), 115–18,
139, 140
in berry fool, 322
foraging in rain, 246
in kompot, 353
Curry, Lynne, 192
cyanide, 95
Cydonia oblonga, 307–14.
See also quince

D

Damrosch, Barbara, 56
de Acosta, José, 131
dehydrators, 20–21
The Dharma Kitchen
(Havranek), 94
Diamond, Jared, 4
Diospyros spp., 277–86. *See also*
persimmons
Diospyros kaki 'Fuyu' (Fuyu
persimmons), 282–83
Diospyros kaki 'Hachiya' (Hachiya
persimmons), 280–81, 286
Diospyros virginiana (American
persimmons), 278–80, 284, 286

dogwood
flowering (*Cornus
florida*), 326
kousa (*Cornus kousa*), 326
drying fruits and berries, 19–21
apples, 36
autumn olives, 62
blueberries, 81
cranberries, 20, 108
currants, 118
elderberries, 122
figs, 133
ground cherries, 164
loquats, 203
for muesli, 358
peaches, 256
pears, 267, 269
persimmons, 281
plums, 290
rose hips, 328
serviceberries, 331
spicebush berries, 335, 337
sumac, 348

E

eastern mayhaw (*Crataegus
aestivalis*), 214
eastern prickly pear (*Opuntia
humifusa*), 301
eastern red cedar (*Juniperus
virginiana*), 177–78, 179
Elaeagnus angustifolia (Russian
olive), 61
Elaeagnus pungens (thorny olive
or silverthorn), 61
Elaeagnus umbellata (autumn
olives), 59–63, 117
elderberries (*Sambucus* spp.),
73, 119–28
in cordial, 123, 124–25
in jelly with sumac-infused
water, 348
in kir royale, 123
in kompot, 353
toxic leaves and stems of,
10, 121
elderflowers, 119, 120–21, 122
Emmons, Didi, 62
Endecott, John, 266

envy, fruit, 193
Eriobotrya japonica, 201–6. *See
also* loquats
Eureka lemons, 183, 184
European bittersweet (*Solanum
dulcamara*), 163

F

fallen fruits and berries
ground cherries, 164
locating source of, 89
mayhaws, 214
mulberries, 223
oranges, 234
pawpaws, 244
pears, 264, 272
false bananas, 241. *See also*
pawpaws
false grapes (*Parthenocissus
quinquefolia*), 159
fats, 26–27
fermentation, 18–19, 20
bokashi, 219
of cranberry relish, 109
of Old Sour, 200
of preserved lemons, 190
supplies for, 20, 24
figs (*Ficus carica*), 13, 129–38, 193
in cake, 134–35, 138
kale with, 137
pork, rosemary, and onions
with, 136–37
preserved, 134, 138
flavor of foraged fruits, 170
flour, 23, 27
buckwheat, 175
rye, 356–57
flowering dogwood (*Cornus
florida*), 326
flowering quince (*Chaenomeles
japonica*), 311
food mills, 21–22
Food Values in Wartime
(Plimmer), 324
fools, berry, 139, 192, 322
foraging
accessibility of, 15
benefits of, 3–4, 12, 67, 343
boom and bust cycles in, 318

prairie rose (*Rosa arkansana*), 323–24

preserves, 51, 212
    currant, 118
    fig, 133, 134, 138
    gooseberry, 142
    grape, 152
    lemon, 187, 190–91
    pawpaw, 246
    pectin in, 113
    plum, 290
    sugar in, 103

preserving fruits, 17–21, 364

pressure cookers, 21

prickly pears (*Opuntia* spp.), 57, 301–6
    in agua fresca, 306
    glochids or spines of, 13, 301, 303

protease enzymes in figs, 133

prune plums, 290

Prunus spp. *See also specific fruits.*
    apricots, 47–54
    cherries, 85–96
    nectarines, 256
    peaches, 253–62
    plums, 287–93

Prunus americana (American plum), 288

Prunus angustifolia Marshall (Chickasaw, sand, or sandhill plums), 288

Prunus armeniaca, 47–54. *See also* apricots

Prunus cerasus (sour cherries), 86–87, 94–95

Prunus maritima (beach plums), 288

Prunus persica, 253–62. *See also* nectarines; peaches

Prunus serotina (black cherries), 11, 87–88, 89, 90

Prunus virginiana (chokecherries), 55, 88–89, 90, 353

puddings
    pawpaw, 249
    persimmon, 286
    rice, 354

Puglisi, Christian, 180

Punica granatum, 295–300. *See also* pomegranates

Pyrus spp., 263–76. *See also* pears

Pyrus communis, 269

Pyrus pyrifolia, 268–69. *See also* Asian pears

Q

quince (*Cydonia oblonga*), 7, 9, 307–14
    in applesauce, 39
    and fruit envy, 193
    in jelly, 308, 310, 312–13
    in membrillo, 100, 308, 310, 314
    pectin in, 113, 308, 310, 314
    poached, 314, 354
    rice pudding with, 354

R

rain, foraging in, 246, 347

Rainier cherry, 86

ramanas rose (*Rosa rugosa*), 323

raspberries (*Rubus* spp.), 13, 71, 221, 315–22
    in berry fool, 322
    black, 9, 315
    in buckwheat buckle, 175
    in cake, 293
    in cobbler, 230
    foraging in rain, 246
    in fruit leather, 318
    in muffins, multigrain multi-berry, 225
    pickled, 84
    in pie, 74
    red, 316, 321, 322
    salmon glazed with, 321
    in sangria, 45, 262
    in shortcakes, 342–43
    in shrubs, 226, 227
    on U-pick farms, 83

red aronia (*Aronia arbutifolia / Photinia pyrifolia*), 56

red cedar, Eastern (*Juniperus virginiana*), 177–78

red currants, 116, 118

red elderberry (*Sambucus racemosa*), 119

red huckleberries (*Vaccinium parvifolium*), 173

red mulberry (*Morus rubra*), 221–22

red raspberries (*Rubus idaeus*), 316, 321, 322

red whortleberry (lowbush cranberry), 106

Reichl, Ruth, 184, 185

relish
    cranberry, 108, 109
    ground cherries in, 164

Rembold, Sean, 259

Resnick, Lynda, 296

restaurants
    foraged ingredients served in, 180
    foraging near parking lots of, 319–20

rhubarb, mulberries with, 224

Rhus spp. (sumac), 345–51

Rhus coriaria (Sicilian sumac), 346

Rhus glabra (smooth sumac), 346, 350

Rhus typhina (staghorn sumac), 346, 350

ribena, 116

Ribes spp. *See also specific berries.*
    currants, 115–18
    gooseberries, 115, 139–46
    and white pine blister rust, 115–16, 139

rice pudding, 354

ricin, 332

Ricinus communis (castor bean), 332

roadside fruits, safety of, 11, 120

Robbins, Tom, 70

Rocky Mountain grape (Oregon grape), 13, 207–10

Roden, Claudia, 49

root vegetables with pomegranate drizzle, 300

Rosa spp. (rose), 323–28

Rosa arkansana (prairie rose), 323–24

Rosa rugosa (ramanas rose), 323

ABOUT THE AUTHOR

Andi Roberts

SARA BIR is a chef and a writer. A graduate of the Culinary Institute of America, she creates recipes that draw on her professional skill set, yet are realistic for home cooks. Bir has worked as a chocolate factory tour guide, sausage-cart lackey, food editor, recipe tester, restaurant critic, librarian, and arts and entertainment reporter. In her spare time, she enjoys walking around and looking at plants. Bir's writing has been featured in *Saveur*, *Edible Ohio Valley*, *Best Food Writing 2014*, two *Full Grown People* anthologies, and on the websites Serious Eats, Lucky Peach, and Paste. She lives in Ohio.

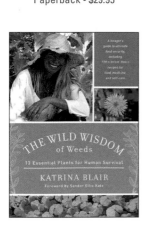

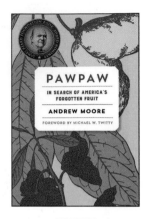
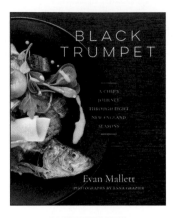

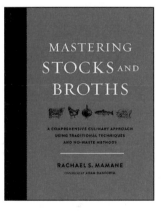